T0258562

Recent Developments in Crystallization

Recent Developments in Crystallization

Edited by **Garry Hollis**

New York

Published by NY Research Press,
23 West, 55th Street, Suite 816,
New York, NY 10019, USA
www.nyresearchpress.com

Recent Developments in Crystallization
Edited by Garry Hollis

International Standard Book Number: 978-1-63238-387-7 (Hardback)

Printed in the United States of America.

Contents

Preface   VII

Fundamentals and Theoretical Aspects   1

Chapter 1   **Crystallization in Glass Forming Substances:
The Chemical Bond Approach**   3
Elena A. Chechetkina

Chapter 2   **Numerical Models of Crystallization and Its
Direction for Metal and Ceramic Materials in
Technical Application**   29
Frantisek Kavicka, Karel Stransky, Jana Dobrovska,
Bohumil Sekanina, Josef Stetina and Jaromir Heger

Chapter 3   **Crystallization Kinetics of Chalcogenide Glasses**   55
Abhay Kumar Singh

Chapter 4   **Recrystallization of Active Pharmaceutical Ingredients**   91
Nicole Stieger and Wilna Liebenberg

Chapter 5   **A Mathematical Model for Single Crystal Cylindrical
Tube Growth by the Edge-Defined Film-Fed
Growth (EFG) Technique**   113
Loredana Tanasie and Stefan Balint

Chapter 6   **Crystallization in Microemulsions:
A Generic Route to Thermodynamic Control
and the Estimation of Critical Nucleus Size**   143
Sharon Cooper, Oliver Cook and Natasha Loines

Chapter 7   **Chemical, Physicochemical and Crystal – Chemical
Aspects of Crystallization from Aqueous
Solutions as a Method of Purification**   171
Marek Smolik

Permissions

List of Contributors

Preface

The purpose of the book is to provide a glimpse into the dynamics and to present opinions and studies of some of the scientists engaged in the development of new ideas in the field from very different standpoints. This book will prove useful to students and researchers owing to its high content quality.

Crystals can be formed from solids, melts, liquids and even vapors. Crystals are of various kinds, both organic and inorganic, and the process of crystallization is an age-old multidimensional subject of research study. Though complex in its characteristics, the process of crystallization forms a part of our regular lives from making ice cubes in our refrigerators to latest and advanced chemical and electronic industry processes. This book is aimed at bringing forth novel perspectives while presenting important data for experts in the related fields of crystallography. It elaborates the fundamentals and theoretical aspects in crystallization. The reader is introduced to various different perspectives of the crystallization process which will prove to be a valuable reference source to those with diverse backgrounds.

At the end, I would like to appreciate all the efforts made by the authors in completing their chapters professionally. I express my deepest gratitude to all of them for contributing to this book by sharing their valuable works. A special thanks to my family and friends for their constant support in this journey.

Editor

Fundamentals and Theoretical Aspects

Crystallization in Glass Forming Substances: The Chemical Bond Approach

Elena A. Chechetkina

Institute of General and Inorganic Chemistry of Russian Academy of Sciences, Moscow, Russia

1. Introduction

Glassy materials are strongly connected with crystallization or - more strictly - with the ability to *avoid* crystallization when cooling a melt. The more stable a supercooled liquid against crystallization, the higher its glass forming ability. It should be noted that almost every substance can be preapared in the form of amorphous solid by means of special methods of fast melt cooling (in the form of ribbons), evaporating (films), deposition (layers) and sol-gel technique (initially porous samples), etc. [1]. The resulting materials are often named "glass" (e.g., "glassy metals"), but they are actually outside the scope of this chapter. Here, we consider only typical inorganic glasses, such as SiO_2 and Se, which correspond to an understanding of glass as a *bulk non-crystalline solid prepared by melt cooling*. "Bulk" means a 3D sample having a size of 1 cm and larger, a condition that implies a cooling rate of about 10 K/s and lower. Despite the "technological" character of this definition, it is the most objective one, being free from the declared or hidden speculations about the nature of glass.

The theoretical background is given in Section 2, beginning with classical notions about crystallization in ordinary melts and the concept of the *critical cooling rate* (CCR), understood as the minimal cooling rate that provides the solidification of a melt without its crystallization. As far as any crystallization ability is reciprocal to a glass forming ability, one can evaluate a *glass forming ability* in terms of the CCR, using both crystallization theory and experiments. It will demonstrate a principal inapplicability of classical crystallization theory in the case of glass forming substances. As an alternative, we have developed a new approach based on the hypothesis of *initial reorientation*, considered as a specific pre-nucleation stage in a non-crystalline network. The classical model of a *continuous random network* consisting of common covalent bonds (two-centre two-electron, 2c-2e) is modified in two respects. First, we introduce *hypervalent bonds* (HVB) in addition to covalent bonds (CB); and besides, such bonds can be transformed into one another (CB↔HVB). Second, the elementary acts of bond exchange are in spatiotemporal correlation, thus providing a *bond wave* by means of which the collective processes - including initial reorientation and further crystallization - may occur. A bond wave also means a non-crystalline order of hierarchical character: from the well known short-range order (it changes in the vicinity of HVB), through the well known but poorly-understood medium-range order (in the limits of the wavefronts populated with HVB), to the non-crystalline long-range order generated by bond waves. Insofar as bond waves change their parameters and direction during melt cooling, being frozen in solid glass, the non-crystalline long-range order depends not only on the substance under consideration but also on the sample prehistory. This is a principal distinction from the ordinary long-range order in crystals, which is strictly determined for a given substance at a given temperature. As such, the problem of crystallization in glass can be considered as a competition between two types of order, and hypervalent bonds and their self-organization in the form of a bond wave play a central role in this process.

In Section 3 and Section 4, original crystallization experiments based on the above approach are described. In Section 3, we consider crystallization in **solid Se-X** glasses, in which additions of a different chemical nature (X = Cl, S, Te, As, Ge) are used. A set of properties, including the abilities of nucleation and surface crystallization, was investigated based upon the composition for each series. A strong *non-linearity* was found in the region of small additions N<N* (concentration N* depends on the nature of X), which is discussed in terms of the bond wave interaction with foreign atoms also existing in a hypervalent state. Finally, in Section 4, original experiments on the crystallization of **softening Se-X** glasses in an **ultrasonic field** are presented. In softening glass, bond waves are refrozen and an ultrasonic field can act as an *information field* for them; in our case, a US-field can give a predominant direction for the crystallization process. Actually, the resultant glasses become *anisotropic*, the anisotropy also being non-linear in respect to the composition.

Thus, starting from the classical approach operating with a 3-step process of crystallization (sub-critical unstable nucleation, stable nucleation, crystal growth), fluctuations, atomic jumps, atomic/geometrical structure, one-type bonding and one-type long-range order, we pass to a strange picture of mixed bonding, bond waves, "wavy" long-range order and glass as a self-organizing system.

The text is intended for a wide audience. The readers from the field of glass are invited to meet with new experiments and new ideas. For beginners – let us now enter into the mysterious world of glass!

2. Crystallization in supercooled glass forming liquids (T>T$_g$)

2.1 Critical cooling rate

The idea of the CCR as a measure of a glass forming ability seems to be trivial, but it remained in the shadows up until 1968, when *Sarjeant & Roy* [2] emphasized "the concept of *'critical quenching rate'*... defined as the cooling rate below which detectable crystalline phases are obtained from the melt" and proposed the first theoretical expression for the CCR in the form of:

$$Q_c = 2.0*10^{-6}*(T_m)^2*R/V*\eta \qquad (1)$$

Here T_m is the melting point, η (poises) is the viscosity at T_m and V is the volume of the "diffusing species". The main problem is the evaluation of V, which needs "an extreme simplification and interaction of many parameters involved" [2]. Nevertheless, it is seen in **Table 1** that the values thus obtained are in agreement with the later estimations of the CCR [3-5] using the equations of classical crystallization kinetics.

Strictly specking, it is easy to fit into the "classical" gap [3-5] that extends for orders in magnitude. More interesting is the tendency for this gap, $\Delta|\lg Q_c|$, to be especially wide just for glass-formers (G). It is seen in **Fig.1** that in glass forming substances ($Q_c<10^2$ K/s by definition) the difference $\Delta|\lg Q_c|$ increases when Q_c decreases, i.e., the larger the gap, the greater the glass forming ability. Moreover, and to the contrary, for non-glass forming substances ($Q_c > 10^2$ K/s), there is no relation between Q_c and its variation, the latter actually being the same for all non-glass formers.

Substance	T_m, K [4]	lgQ_c(K/s)				$|\Delta lgQ_c|$
		[2]	[3]	[4]	[5]	
SiO$_2$ (G)	1993	+0,6	-3,7	-1	-2,9	4,3 (G)
B$_2$O$_3$ (G)	733		-1,2	-16	-7,3	8,7 (G)
GeO$_2$ (G)	1359			-11	-2,9	9,8 (G)
P$_2$O$_5$ (G)	853			-23	-5,9	17,1 (G)
As$_2$O$_3$	1070			+7	+4,1	2,9
H$_2$O	273	+7,2	+7	+12	+10,4	5,0
Salol (G)	316.6		+1,7	-15		13,3 (G)
Ethanol	159			-2	+1	3,0
Glycerin (G)	293			-40	-7,9	32,1 (G)
CCl$_4$	250,2			0	+5,2	5,2
C$_6$H$_6$ (G?)	278,4			-14	+4,9	18,9 (G?)
BeF$_2$ (G)	1070			-3	-4,0	1,0 (G)
ZnCl$_2$ (G)	548			-21	-4,4	16,6 (G)
BiCl$_3$	505			+7	+5,5	1,5
PbBr$_2$	643			+3	+6,3	3,3
NaCl	1073	+9,1		+10	+8,4	1,6
NaNO$_3$	580			+5	+7,0	2,0
NaCO$_3$	1124			+3	+5,0	2,0
NaSO$_4$	973			+10	+7,4	2,6
NaWO$_4$	1162			+7	+8,4	1,6
Ag	1234	+8,0	+10	+8	+9,3	1,3
Cu	1356			+9	+8,5	0,5
Sn	505,7			+2	+7,9	5,9
Pb	600,7			+10	+8,7	1,3

Table 1. The calculated CCRs among the groups of oxides, substances with hydrogen bonding and/or organic glass formers, halides, sodium salts and metals. The "G" in brackets indicates a well-recognized glass former.

One should note the absence of benzene (see C$_6$H$_6$ in Table 1) in Fig.1. This is a result of the outstanding disagreement in the evaluations of its CCR: benzene looks like an excellent glass former after [4], but after [5] it is a non-glass former. When lgQ_c=+4,9 after [5], the point for difference $\Delta|lgQ_c|$=18,9 is dramatically lifted above the "amorphous line"; however, if one takes lgQ_c=−14 after [4], the point falls far below the correlation line for glass formers.

The Qc interval in Fig.1 extends for 19 orders in magnitude. The practical interval is far narrower. For glasses, there are three useful CCR points: the glass definition limit Q_c=10^2 K/s, which corresponds roughly to a quenching of a 1 cm^3 sample in water with ice, the rate of 10 K/s corresponding to quenching in the air, and the rate of 1-10^{-2} K/s corresponding to free cooling in a furnace. Therefore, the extremely large negative values of lgQ_c(K/s) in Table 1 - such as -16 for B$_2$O$_3$ or -23 for P$_2$O$_5$ [4] - only display an extremely high glass forming ability, i.e., the ability for very slow cooling in a furnace, a feature which provides the opportunity to obtain very massive articles, such as lenses for telescopes. For amorphous materials ($Q_c > 10^2$ K/s), the experimental limit is about $Q_c = 10^6$ K/s, which corresponds with the preparation of amorphous ribbons, namely the 2D case. The calculated value of $Q_c = 10^8$-10^{10} K/s displays only an extra low amorphization ability of a substance, which cannot be obtained by melt cooling at all. This is really the 1D case, when the amorphous film is prepared by evaporation onto a cold substrate.

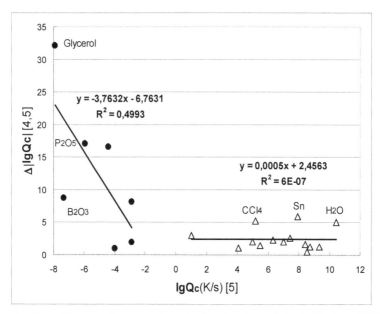

Fig. 1. The absolute difference between the calculated values of the CCRs from [4,5] as a function of the CCR value after [5]. The data is from Table 1.

Based on the tendency in Fig.1, one can conclude that the greater the glass forming ability, the more problematic the application of the classical theory of crystallization. In reality, in the case of a lithium disilicate glass it was shown that "not only do all forms of classical theories predict nucleation rates many orders in magnitude smaller than those observed, but also the temperature dependence of the theoretical rate is quite different from that observed" [6]. Thus, crystallization in glasses needs special methods, both theoretical and experimental.

Although there are many experimental works on the crystallization of glass and glass forming liquids, only a few of them concern the CCR. The known evaluations of Q_c in oxides [7-11] and chalcogenides [12-16] are made by either a direct method covering a relatively small $\lg Q_c$ interval (e.g., from -2 to +4 after [7-10]), or indirect methods. By 'indirect' is meant that the crystallization data corresponding to some extent of crystallinity $\alpha \sim 0.01$-0.1 is recalculated to a much lower α corresponding to Q_c (usually to $\alpha = 10^{-6}$ [15-16]) using equations of the *Kolmogorov-Avrami* type. However, the applicability of such equations in an extra low α region is also under question. Thus, it is not surprising that interest in the investigation of the CCR is low at present. The reader can find a critical review of the "golden age" of the CCR in the last third of the Twentieth Century in our monograph [17]. The main result of this period seems to be the formulation of new problems concerning glass forming ability.

The general question is what is meant by the demand of melt cooling "without crystallization"? After *Vreeswijk et al.* [4] glass is considered as being partially crystallized even one critical nucleus is formed; this is a **theoretical** limit for estimation of the Q_c value. Later, *Uhlmann* [3] proposed the crystalline fraction α=10⁻⁶ for the **experimentally** detected limit, and *Ruscenstein & Ihm* [5] proposed α=10⁻⁴ for a more convenient **practical** limit. Obviously, the lower α, the higher Q_c for a given substance. However, when comparing the calculation results in Table 1 - we see that there is no a relation between the crystalline fraction permitted (α increases from [4] to [5]) and the corresponding Q_c values. Thus, classical crystallization theory needs serious reconsideration for glass forming substances.

2.2 Induction period, transient effects and initial reorientation

The critical nucleus - arising when overcoming the "thermodynamic" barrier W* with the following steady-state nucleation of the *I(T)* frequency and the steady-state crystal growth of the *u(T)* rate - forms the basis of classical crystallization theory. The critical nucleus is considered as the cause of the *induction period* observed in the process of crystallization. It is unclear, however, to what extent the classical notions are true for glass forming substances. In order to test the assumption about steady-state nucleation, *Kelton & Greer* [18] have analysed the data for lithium disilicate (a typical glass former) and two metallic "glasses" ('amorphous metal' would be a more appropriate term). When the authors introduced the time-dependent nucleation frequency *I(τ,T)*, they revealed the so-called *transient effects* which occurred to be most significant only for typical glass. Transient effects, together with the fact that *"even small uncertainties of material parameters can introduce uncertainties of several orders in magnitude in the calculated nucleation frequency"* [18], make the application of classical crystallization theory problematic in the case of glass forming substances.

On the other hand, from the beginning of the 1980s onwards, we have developed an alternative approach for the prediction of glass forming ability in relation to crystallization ability. This approach is based on *Dembovsky's* "empirical theory of glass formation" [19], by which one can calculate the dimensionless glass forming ability, which was transformed further into an energetic barrier for crystallization E_{cr} [20]. The next stage was the transformation of the semi-empirical E_{cr} (enthalpy) to the barrier $\Delta G^{\#}$ (free energy), and finally $\Delta G^{\#}$ to the CCR measure V_m [21,22]. The reader can find a complete calculation scheme for V_m in [17; p.145] or [23; p.105]. Two points of this CCR are of special interest.

First, in order to calculate V_m, one needs *no kinetic parameters*, such as a coefficient of diffusion and/or viscosity. Instead, we use only the phase diagram of a system (in order to determine the liquidus temperature for a related unit: an element, a compound, an eutectic or else an intermediate), the first coordination number (presumably, in the liquid state at around T_m) and the averaged valence electron concentration (e.g., 6 for Se, 5.6 for As_2Se_3, 5.33 for SiO_2, etc.).

Second, the barrier $\Delta G^{\#}$ was attributed to a special pre-crystalline ordering named the *initial reorientation* (IReO), which is postulated as the stage after which crystal nucleation and growth is possible [21]. IReO means that neither nucleation nor crystal growth can proceed in a non-oriented - i.e., a truly "amorphous" - medium. Since IReO is a simple activation process with the $\Delta G^{\#}$ barrier, the reoriented fraction increases exponentially with time, as the dotted line in Fig.2 shows.

States:

(1) $A \to A_0 \to A_{Nu}$

(2) $A_0 \to A_{Cr}$

(3) $A \to A_0 \to A_{Cr}$

Barriers:

(1) $\Delta G^{\#}$; $(W^* + \Delta G^{\#})$

(2) $\Delta G' \to \Delta G''$

(3) $\Delta G^{\#}$; $\Delta G''$

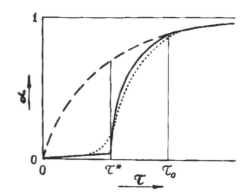

Fig. 2. Isothermal crystallization after [21]: the dotted curve describes the development of the reoriented non-crystalline fraction; the solid line is the model temporal dependence of the extent of crystallinity α; the pointed curve is a real crystallization process.

Atomic states: A is the initial truly "amorphous" state for a species in a supercooled liquid or glass; A_0 is the reoriented state; A_{Nu} is the state of the crystal nucleus; A_{Cr} is the state of the growing crystal.

Barriers: $\Delta G^{\#}$ is the initial orientation; W^* is the critical nucleus; $\Delta G'$ is the steady-state nucleation; $\Delta G''$ is the crystal growth.

The three intervals in Fig.2 correspond to an *induction period* at $0 < \tau < \tau^*$, a *transient period* at $\tau^* < \tau < \tau_0$ and a *saturation* at $\tau > \tau_0$. At first, the reoriented areas are relatively small and/or disconnected, such that only nucleation can proceed. Although nucleation is permitted, the extent of the crystalline fraction (the solid line) is negligible when compared with the reoriented fraction (the dotted line). At the moment τ^*, the reoriented areas become large enough and/or percolated so as to permit not only nucleation but also crystal growth, using both the arising and newly existing portions of the reoriented species. In this *transient* region $\tau^* < \tau < \tau_0$, crystallization slows gradually and at τ_0 it becomes limited by the process of initial reorientation when each new portion of the reoriented species joins directly to the growing crystals.

One should note in Fig.2 the well known barriers G', G'' and W^*. Fortunately, the problems with these barriers (e.g., varying G' after [18] or an anomalous increase of W^* with a decreasing temperature at about T_g [24]) are outside of our method for the estimation of the CCR since V_m relates only to the initial reorientation stage with the $\Delta G^{\#}$ barrier which determines the boundary between glassy and crystalline states. Thus, V_m is the *upper* theoretical limit for the CCR and in this sense V_m satisfies the condition set by *Vreeswijk et al.* [4] concerning the absence of even one critical nucleus when determining the CCR.

By 'initial reorientation' is meant some kind of order. Interestingly, the notions concerning pre-crystalline ordering in glasses appear consistently in the literature. One can see the IReO in a real SEM image, which is interpreted by the authors as "a pre-crystallization stage, in which glass matrix becomes inhomogeneous, forming nano-sized volumes" [25]. Indirect evidence of the IReO is seen in the conclusion that "glass-transition kinetics can be treated as pre-crystallization kinetics" [26] as well as in the models of "dynamic heterogeneity" in an amorphous matrix before its crystallization [27-29]. The problem is that both "dynamic heterogeneity" and "initial reorientation" are the only terms that need decoding.

2.3 Glass structure, hypervalent bonds and bond waves

The theory of glass structure was actually formed at 1932 when, beginning with the words *"It must be frankly admitted that we know practically nothing about the atomic arrangement in glasses,"* the young scientist *Zachariasen* proposed his famous model of a **continuous random network** (CRN) [30]. This network consists of covalent bonds, - the same as that in a related crystal - and the only difference is the "random" arrangement of the bonds in CRN in contrast with their regular arrangement in a crystalline network. As far as the atomic arrangement in CRN is determined by the arrangement of chemical bonds, one might suppose that only chemists - who deal with chemical bonds - would play a leading role in the further development of the CRN model. However, the development was directed in another "physical" way. As a result, the chemical bond is present in contemporary glass theory in the form of "frustration", "elastic force" and rigid "sticks" connecting atoms-balls, etc., but not as a real object with its own specificity. We think that a pure "physical approach" opens a rich field for theoretical speculation, rather than clarifying the nature of glass. For example, to understand glass transition, the analogues with spin glasses, granular systems and colloids are used (e.g., see the excellent review [31]), although any similarity in behaviour does not mean that the similarity is the reason behind that behaviour.

One can find, however, deviations from the main "physical" stream of glass science, even among physicians. As a fresh example, we may cite that: *"It is widely believed that crystallization in three dimensions is primary controlled by positional ordering, and not by bond orientational ordering. In other words, bond orientational ordering is usually considered to be merely a consequence of positional ordering and thus has often been ignored. ... Here we proposed that bond orientational ordering can play a key role in (i) crystallization, (ii) the ordering to quasi-crystal and (iii) vitrification"* [29]. Although this is a rare case, it may be a sign that the need for a chemical approach is now stronger.

Our goal was to return the chemical bond - as chemists know it - to the theory of glass. As far back as in 1981, *Dembovsky* [32] has connected the experimental fact of an *increased* coordination number in glass forming melts with the model of *quasimolecular defects* (QMD) following *Popov* [33] for the creation of the chemical-bond basis for glass theory. The principal feature of a QMD is its hyper-coordinated nature, a property that provides for the connectivity of a covalent network even at melt temperature, thus resulting in high melt viscosity, which is a characteristic property of a glass forming melt. QMDs in a non-crystalline network can also explain general features of glass [34-38]. The problem is that the concentration of QMDs should be high enough, especially in a glass forming liquid, to provide these features. Therefore, we have quickly substituted the initial term QMD (quasimolecular *defect*) for TCB (three-centre *bond*) and then, being based on a special quantum-chemical study (see the reviews [39] for chalcogenide glasses and [40] for oxide ones), TCB for HVB, **hypervalent bond.** TCB is only a particular case of HVB, which is rarely realizes in glass. For example, in the simplest case of Se not TCB, having two three-coordinated atoms (-Se<) and one "normal" two-coordinated atom (-Se-) [33], but a four-coordinated atom (>Se<)was revealed [41]. Thus, we have introduced *alternative bonds* into CRN, the bonds whose concentration is commensurable with that of ordinary covalent bonds constituting classical CRN.

The term "hypervalent", which was introduced by *Muscher* in 1969 [42], has a long and controversial history, beginning from the 1920s up until the present (see [43] for an introduction). Currently, a large number of hypervalent molecules is known, and various methods of their theoretical description exist. For a long time, one of the most popular was the *Pimentel's* model [44] of the electron-rich three-centre four-electron (3c-4e) bond; this model was used by *Popov* to construct his "quasimolecular defect" in a covalent network of Se glass [33]. The principal step made by *Dembovsky* was in the understanding of QMD not as a "defect" but as the second type of bonding in a glassy network, the first being a common two-centre two-electron (2c-2e) covalent bond. The next step was made by means of *ab initio* quantum-chemical modelling, which reveals various metastable hypervalent configurations - configurations in which a central atom has more bonds with its neighbours than the "normal" surrounding covalent species - in glasses. We use the term *hypervalent bonds* in order to emphasize the additional bonding state in non-classical CRN, and so the additional structure possibilities in it. This network is not "random" now.

Three types of order can be realized in such a mixed network. When the diffraction pattern of glass is transformed into a radial distribution function one can observe the so-called *short-range order* (SRO) extending to the limits of at least two coordination spheres around an arbitrary atom. The peak's position relates to the distances between the nearest neighbours and the peak square to the number of these neighbours. The SRO in glass is close (but not coinciding completely!) with the SRO in a counterpart crystal, with both SROs being determined by a bond length (first distance), valence (first coordination number) and valence angles (second coordination sphere). The SRO is the basis for a conventional continuous random network, but the problem is that a CRN consisting of covalent bonds cannot exist because of the rigid and directional character of the covalent bond. When moving from the first atom, the stress due to bond distortion accumulates rapidly up to a critical value above which the covalent bond should be destroyed (the covalent bond limits are known in the chemistry of related compounds). This means that a real CRN should contain either internal fractures (however, glass is known to be an optically transparent and mechanical stable material) or additional "soft" regions for relaxation. These regions are provided by HVB and are soft and flexible when compared with covalent bonds.

The next step of order which surely exists in a non-crystalline state is the so called *medium-range order* (MRO), which is observed directly in the diffraction pattern of glass or glass forming liquids in the form of the well known *first sharp diffraction peak* (FSDP). It is a very narrow (for an amorphous state) peak located at about Q_1=1-1,5Å^{-1}; the peak position and intensity depend upon the chemical composition, temperature and pressure (see [45,46] for a review). The two generally accepted parameters of the MRO are the correlation length $d=2\pi/Q_1\approx$6-4Å and the coherence length $L=2\pi/\Delta Q_1\approx$10-20Å, where ΔQ_1 is the FSDP half-width.

The nature of the FSDP/MRO is debatable at present. Our interpretation of the MRO is based on hypervalent bonds and their collective behaviour in the form of a *bond wave*, which is illustrated in **Fig.3**.

A bond wave is the spatiotemporal correlation of elementary acts of the reversible transformation of a covalent bond and a hypervalent bond: CB↔HVB. A snapshot of a bond wave, which spreads in the Se-like network, is shown in the left part of Fig.3. The wavefronts represent equidistant layers populated with hypervalent bonds (they are modelled by TCB in Fig.3), and so the correlation length $d=2\pi/Q_1$ roughly corresponds with the HVB length. The layers give a true Bragg reflex at $Q_1=2\pi/d$, so the FSDP intensity and the FSDP width depend upon the number of these reflecting layers and their reflection ability, with both depending on the temperature.

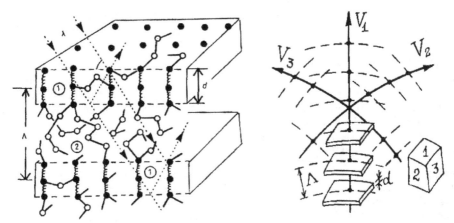

Fig. 3. Bond waves after [46]. On the left: two adjacent wavefronts – layers "1" populated with a HVB (here modelled by a TCB: black atoms linked by spring) together with a CRN "2" between the layers (here an Se-like CRN, which consists of white atoms, each having two covalent bonds as shown by the lines). On the right: the intersection of three bond waves, which gives the 1-2-3 elementary cell.

A bond wave with a Λ wavelength represents a totality of d-layers, and the totality represents a periodic structure of the Λ period. Thus, there appears a Λ-lattice and a corresponding *long-range order* (LRO) in glass. This order is, in some respects, similar to the smectic-type LRO; however, in glass - in contrast with liquid crystal - there are two types of bonding species in the layer and the d-layers themselves are punctuated with thick pieces of the CRN. Such long-range periodicity is "invisible" to ordinary X-ray analysis, by which one can see a CRN (in the form of the peaks in the radial distribution function) and d-layers (in the form of a FSDP in an initial diffraction pattern), but not the Λ-lattice. The reflex from the Λ-lattice should be disposed in the 0.01-0.001Å^{-1} range, which is inaccessible to the usual X-ray techniques. Thus, special techniques (e.g., those using synchrotron radiation) are needed to see a LRO in glass.

One can note that the layer structure shown in the left side of Fig.3 is *anisotropic*, while glass is known to be an *isotropic* medium. Note, however, that anisotropic glasses can also be prepared (our own experiments of this sort will be considered below). The generally observed isotropic behaviour of glass means the *solitonic* behaviour of bond waves, i.e., the waves ability to intersect each other without distortion, as is shown on the right in Fig.3. One wave propagating through a CRN with the velocity V_1 forms the layer structure, two waves (V_1+V_2) create the columnar structure, and three waves ($V_1+V_2+V_3$) correspond to the cellular structure, which is isotropic on the macroscopic scale. The factors which govern bond waves and, consequently, the structure that they form, can be divided into internal (chemical composition, temperature, pressure) and external (flows of energy and/or information) categories.

As far as the elementary act CB$\leftrightarrow$HVB is considered to be a thermally activated process, both the wave frequency and the HVB concentration increase with temperature, while the wavelength Λ decreases [47]. Thus, when approaching the glass transition temperature T_g, the interlayer distance Λ becomes so large that the correlation between the layers/wavefronts becomes impossible, although the intimated HVBs within the layers continue to "feel" each other. This means that a 3D bond wave (Fig.3, right side) stops its propagation through the structure and the 2D bond wave within the stopped d-layers (Fig.3, left side) remains mobile. Thus, the glass transition can be considered as a 3D→2D bond wave transition [48].

As far HVBs represent active sites in the mixed CB/HVB network, one should distinguish between the processes above and below T_g. More specifically, the abovementioned pre-nucleation stage in the form of initial reorientation (IReO) proceeds within the stopped d-layers below T_g and only after the reconstruction of the layers can it penetrate into the CRN: the process is slow and has an induction period (see $\tau < \tau_0$ in Fig.2). Above T_g, where the 3D bond waves exist, the d-layers pass through every structure element: the induction period is short, if it even exists, and the process is much faster and homogeneous.

Crystallization is not the first property considered by means of the bond wave model. Earlier, this model was applied successively for the interpretation of the thermodynamic features of glass forming substances [49], characteristic glass fractures [50], the first sharp diffraction peaks [46] and the temperature dependence of viscosity [51]. Unfortunately, there are only interpretations and so far there is no direct evidence of bond waves at present, i.e., the direct observation of a Λ-lattice in a structural experiment. Instead, we performed a computational experiment [52], presented in **Fig.4**, in which we intended to investigate the ability of the HVB for association – this property alone is a necessary requirement for the existence of a bond wave.

In model clusters like those shown in Fig.4, a single HVB looks like a "defect" embedded into an ordinary continuous random network (CRN). A single HVB was shown to be a low-energy "defect" (compare 0,3 eV for Se_4^0 [41], with 2 eV needed to generate a broken bond $2Se_2 \rightarrow 2Se_1$ or with 1 eV proposed for the so called "valence alternation pair" $Se_3^+Se_1^-$ after [53], the index below is the coordination and the index above is the charge). Nevertheless, even such low-energy defects cannot ensure the above-mentioned over-coordination in glass (there needs to be ~1% for four-coordinated atoms) and especially so in a melt (ten%) [37]. From Fig.4, it follows that the associated HVBs are much more stable, so that even *negative energy* regions can arise in a CRN for a definite HVB arrangement (SS).

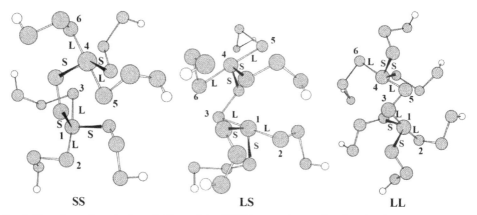

Fig. 4. Quantum-chemical modelling of the HVB interaction in Se after *Zyubin & Dembovsky* [52]. The cluster energies are **−0.16 eV** (SS), +0.08 eV (LS) and +0.09 eV (LL); the notations correspond with the mutual orientation of short (S) and long (L) pairs of bonds surrounding each Se_4^0 atom (1 or 4). The white spheres are the hydrogen atoms terminating the clusters.

In this way, not only is a principal ability to form a bond wave justified, so is the connection of a *structural order* due to a Λ-lattice with an *"energetic order"*, which is based on the intuitive belief that the more structurally ordered substance has the lower energy (with other conditions being equal). As such, the processes of the crystallization of glass, both below and above T_g, can be considered as the competition of two types of order: a crystalline long-range order and a specific non-crystalline long-range order, provided by hypervalent bonds and bond waves. In what follows, these notions are tested by means of special crystallization experiments.

3. Crystallization in solid glass (T<T$_g$)

3.1 The composition dependent rate of nucleation

For these experiments, we chose selenium as the simplest one-element glass; the additions, with various valence abilities, were introduced into the Se matrix, giving five series of Se-X (X = S, Te, As, Ge, Cl) with a varied but relatively low concentration of the second component. The as-prepared samples were cylinders measuring 25 mm in diameter and 15 mm in height; the two cylinder ends were polished. The optical transmission and ultrasonic velocity were measured through the ends; the X-ray fluorescent spectra were measured from the end surface.

The as-prepared samples were quite transparent, actually having the same value of transmission at 1000 cm^{-1} (this value, which corresponds with entry into the so-called "window of transparency" for selenide glasses, we shall call *transparency*) of around 60%, a value that is typical for chalcogenide glasses of high quality for the given thickness of 15 mm. The two other properties (V and r) investigated in the fresh glasses, however, were strongly dependent upon composition, as is seen in **Fig.5**. Note that the both properties are *macroscopic* in character. The ultrasound velocity, V, characterizes the elastic ability of the Se-based network. The relative intensity of the X-ray fluorescence, $r=S_{val}/S_{char}$ (S_{val} and S_{char} are the integral intensities of the $K_{\beta2}$ and $K_{\beta2}$ emission lines for Se, the first corresponding to the $4p\rightarrow1s$ transition from a valence band and the second to the characteristic $3p\rightarrow1s$ inner transition for Se), belongs to the totality of selenium atoms, reflecting the average valence state for Se.

We see a strong non-linear character for all of the dependencies in Fig.5. The two upper cases of Se-Te and Se-S seem to be the most surprising because S and Te belong to the same VI group of the Periodic Table, having the same number of covalently bonded neighbours per atom. Additionally, the Se-Te phase diagram is a simple "fish", which corresponds with the discontinuous series of liquid and solid solutions. However, the *metastable phase diagram* (remember that glass is formed far below the melting point from a metastable liquid) has a "two-fish" form, with "two series of solid solutions meeting at 96.8 at%Se [i.e., 3.2 at%Te – ECh] and temperature 180°C" [55].

Note that the Se-Te glasses display extrema in the 1-4%Te range for the other composition-property dependences, e.g., for electrical and crystallization properties [56] and for the glass transition temperature [57]. *Dembovsky et al.* [58] have also revealed non-linearity when investigating crystallization kinetics in Se-Te glasses in the 0-5% range. In **Fig.6**, the data for CCRs calculated with the use of the data obtained is shown; one can see not only the conditional character of the CCR, the composition dependence of which depends on the chosen α, but also the non-linearity of $Q(N)$ for every chosen extent of crystallinity with the extrema located in the region 1-5%Te. Note that our calculation of the CCR in the form of V_m

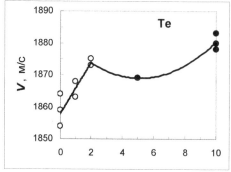

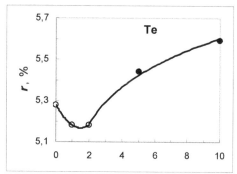

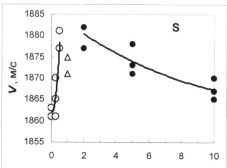

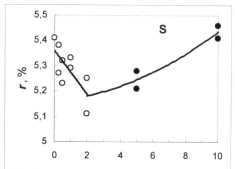

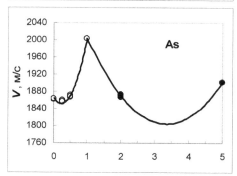

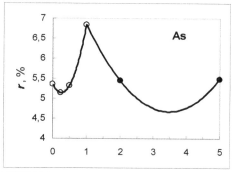

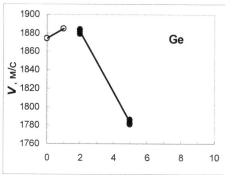

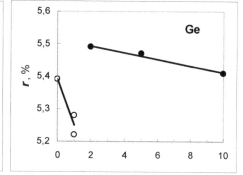

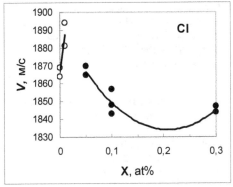
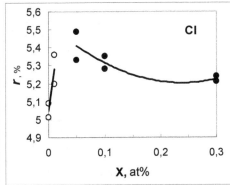

Fig. 5. Longitudinal ultrasound velocity (V) and relative intensity of X-ray fluorescence from the Se valence band (r) in the as-prepared Se-X glasses after [54].

[21,22], which corresponds to the pre-crystallization IReO stage, really represents the upper limit for the CCR, but only outside the non-linearity range. Remember, however, that when calculating V_m we only take into account the *equilibrium* fish-like phase diagram, whereas real glass relates more to metastable diagram(s) with peculiarities at about 5%Te.

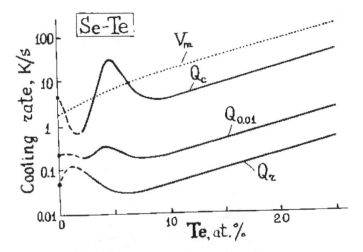

Fig. 6. Critical cooling rates (CCRs) in the Se-Te system after [58]. Q_r, $Q_{0.01}$ and Q_c correspond to the crystalline fraction $\alpha=0,63$; 0,01 and 10^{-6} respectively. V_m is our calculation after [21,22], without the use of crystallization data.

The Se-S series in Fig.5 looks similar to Se-Te, although one can propose a more complex behaviour below 2%S, which can be compared with a more complex equilibrium Se-S phase diagram [59] in contrast with the simple "fish" for Se-Te.

In the Se-As and Se-Ge series, the additions belong to other groups of Periodic Table (IV for Ge and V for As), so one can expect another property-composition behaviour. Accordingly, the $V(N)$ and $r(N)$ dependencies change in the same way in Se-As glasses (maximum on the both), in contrast with the behaviour in Se-Te and Se-S glasses, which display a maximum on $V(N)$ and a minimum on $r(N)$. Although Se-Ge glasses are similar to Se-Te as concerns the maximum on $V(N)$ at 2%Ge and the minimum on $r(N)$ at about 1%Ge, the extrema in Se-Ge are much sharper and there is different post-extreme behaviour in Se-Ge as compared with Se-S and Se-Te.

The Se-Cl glasses are of a special interest because, in contrast with the above four additions which belong to the well recognized glass forming Se-X systems, chlorine in *not* a glass forming addition for Se. It is usually assumed that Cl can only break the Se-Se bonds, thereby decreasing viscosity and thus the glass forming ability. In fact, Se-Cl glasses can be only prepared in the 0-0.3%Cl range, but in other respects the Se-Cl series demonstrates the same behaviour: the as-prepared Se-Cl glasses are quite transparent (about 60%) and their properties are strongly non-linear (see Fig 5, at the bottom).

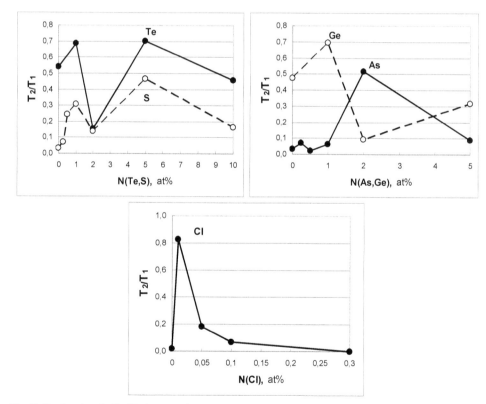

Fig. 7. Darkening in Se-X glasses after 5-year ageing at room temperature [60]. Here, T is the transparency (optical transmission at 1000 cm^{-1}); index "1" refers to the as-prepared glass and "2" to the aged glass. The unchanged transparency, i.e., the absence of darkening, corresponds to $(T_2/T_1)=1$ and complete darkening to $(T_2/T_1)=0$.

Next, the five series of glasses were stored for 5 years at room temperature, which is somewhat below the glass transition temperature ($T_g{\approx}35°C$ for Se). In accordance with the expected crystallization, the **aged glasses** became less transparent, but not to the same extent. It can be seen in **Fig 7** that the transparency shows a non-linear character, like that shown in Fig.5 for ultrasound velocity and electron emission from the valence Se band. Thus, one can suppose that even in fresh Se-X glasses the special regions of initial reorientation (IReO) for further crystallization have been developed. These regions manifested themselves in $V(N)$ and $r(N)$ in Fig.5, but not in the transparency because the IReO regions were congruently inserted into a glassy matrix. During the process of ageing, the IReO regions are transformed into crystalline regions and so become visible due to light scattering from new crystal/glass boundaries.

The question is: what stage of crystallization do we observe in the aged glasses? The X-ray pattern of the aged samples usually looked like a wide hill of a low intensity; only the darkest samples display very weak crystalline peaks on the hill. Thus, because of the low extent of crystallinity (usually lower than 1%) together with the considerable darkening, we can conclude that by the optical transmission method we can observe the *nucleation stage* of crystallization. Next, from Fig.7 one can conclude that compositions of 1%Te and 5%Te, 5%S, 1%Ge, 2%As and (surprisingly!) 0.01%Cl are the most stable against nucleation.

3.2 The composition dependent extent of heterogeneity

Nucleation in glass is known to be usually *heterogeneous*, i.e., the nuclei tend to appear on the surface of a sample rather than distribute evenly in the volume. Note that only the case of homogeneous nucleation is considered in classical crystallization kinetics, a fact that creates additional problems when comparing theory with experiment for such non-classical objects like glass.

In order to evaluate heterogeneity in a numerical way, we have elaborated a simple method which includes the removal of the surface layer [61]. The main effect of the removal is the resection of the surface nuclei with the consequent rise of transparency. Next, the extent of heterogeneity G can be evaluated as:

$$G = (T_3-T_2)/T_1 , \qquad (2)$$

where T_1 is transparency (i.e., the optical transmission at 1000 cm^{-1}) for the as-prepared glass (of 15 mm thickness), T_2 is the transparency after ageing and T_3 is the transparency of the aged sample after the removal of the surface layer (0.5 mm from each side of the cylinder, a procedure that gives a sample of 14 mm thickness). The method is illustrated on the Se-Te series as an example in the left-top and left-bottom (Te) of Fig.8.

From Fig.8, it can be seen that the distribution of crystalline nuclei is most homogeneous in the Se-Ge system having G=0-0.2. The negative G for the 5%Ge sample may indicate the "inverse heterogeneity", a situation whereby nuclei prefer to burn within the volume rather than at the surface. The opposite deviation G>1, which is observed in the Se-Cl series for the 0.05%Cl sample when the volume becomes more transparent after ageing ($T_3>T_1$), may be a result of the sample thinning. Another possibility, of *non-crystalline ordering* during ageing, seems to be more interesting. The reason for this ordering may be the development of the IReO regions in the glassy matrix (see the dotted line in Fig.2) without nucleation.

One can note a remarkable similarity in the $r(N)$ graphs in Fig.5 for the as-prepared glasses and the $G(N)$ graphs in Fig.8 for the aged glasses, and not only in the extrema location, which coincides for every Se-X series, but also in the dependence shape which is qualitatively the same for the Se-Te, Se-As and Se-Cl series. For the Se-S and Se-Ge series, the shapes are somewhat different. One should note, however, that the Se-S glass is a special case because of the introduction of sulphur with T_g= -25°C, which would decrease the glass transition temperature in the series. Accordingly, Se-S glasses are especially inclined to crystallization when ageing at room temperature.

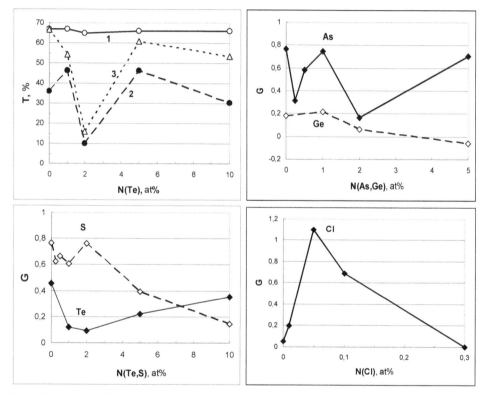

Fig. 8. The extent of heterogeneity, G by eq.(2), for spontaneous nucleation in Se-X glasses. T_i is the optical transmission at 1000 cm^{-1} for the as-prepared glass (1), the aged glass (2) and the aged glass without a surface layer (3).

In spite of these deviations, a non-trivial relation between the inclination to heterogeneous nucleation, $G(N)$, and a change in chemical bonding, reflecting by $r(N)$, is observed. Note that by means of the X-ray fluorescence method we probe only the *surface* layer of the sample; therefore, there exists a direct connection between *nucleation ability* and the *density of valence electrons*, with which the value of $r(N)$ can be related. This is a direct route to hypervalent electron-rich bonds, which were discussed in section 2.3: the higher the electron density, the easier the formation of such HVBs representing active centres in covalent network, the centres which provide for processes including crystallization. The concentration of HVBs may be high owing to the decrease of energy when HVBs are associated (see Fig.4), probably in the form of bond waves. Thus, foreign atoms can interact strongly and non-linearly with the bond waves which spread in non-crystalline network during the glass preparation (3D waves) and ageing (2D waves). What is a mechanism for such an interaction?

3.3 Non-linearity and self-organization in glass

From a phenomenological point of view, one can distinguish between "*impurity*" (non-linear) and "*component*" (smooth) concentration regions [54]. The boundary concentration, after which non-linearity begins to vanish, depends upon the system; it is about 2% for Se-S, Se-Te and Se-Ge glasses, about 1% for Se-As glasses, and about 0.05% for Se-Cl glasses (Figs.5-8). Let us discuss a possible nature of non-linearity in a low-concentration "impurity" region.

Starting from classical CRN [30], one cannot wait for any singularities except for those stipulated by a phase diagram, and it is generally accepted that foreign atoms enter into a CRN with "normal" valence (e.g., Te forms the –Se-Se-Te-Se-Se- chains), thus causing a proportional change in glass properties. The above-demonstrated non-linearity means that this is not the case and, hence, that the valence might be unusual. We have a model of *hypervalent bonds* - HVBs - after Dembovsky (section 2.3) for just this situation. It is clear, however, that the HVBs in a concentration of about 1% cannot explain the *macroscopic* effects observed. This is the *bond wave* model for the case of the collective interaction between "normal" bonds and HVBs. Next, the question about the nature of non-linearity should be reformulated: how do selenium bond waves interact with foreign atoms in a selenium network? Let us begin with the quantum-chemical modelling of the simplest case of Se-Te.

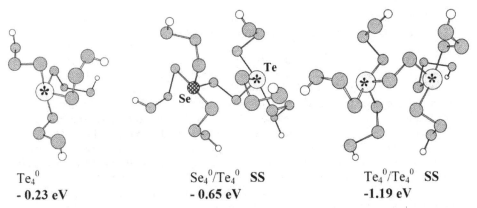

Te_4^0

- 0.23 eV

Se_4^0/Te_4^0 **SS**

- 0.65 eV

Te_4^0/Te_4^0 **SS**

-1.19 eV

Fig. 9. A quantum-chemical modelling of HVBs in a selenium clusters containing Te atoms after [62]. The energies below are compared with the selenium CRN of the same size.

It can be seen in **Fig.9** that even a simple hypervalent Te_4^0 decreases the energy of the surrounding network (-0.23 eV). When the selenium wavefront (see d-layer in Fig.3, on the left) moves closer, a more stable $Se_4^0Te_4^0$ (-0.65 eV) is formed. Hence, a hypervalent Te atom represents an energetic trap for the selenium bond wave; however, this is so for only that local part of the wavefront, a part that interacts with the atom. When the Te concentration is high enough, the atoms can form hypervalent associates with the additional decrease in energy - see dimer $2(Te_4^0)$ in Fig.9.

The result of such an interaction is currently unclear. Given only the traps, foreign HVBs - such as Te_4^0 - should slow down or even destroy a bond wave, most probably to the extent that it is proportional to their concentration. However, we observe both maxima and minima in concentration dependencies in different Se-X series (Figs.5-8); hence, the interaction is not simple and depends on the concentration of foreign atoms. The reasons may be both in their bonding state and/or their arrangement – random or ordered. An ordered state can be realized by a directed diffusion of impurity under the action of any bond waves spreading through the melt, before the waves are frozen in solid glass.

The nature of the added atoms is also significant for their interaction with the bond wave. For example, a similar quantum chemical study for the Se-Cl system [63] leads to the same qualitative result as for with Se [52] and Se-Te [62], namely, the aggregation of HVBs with the participation of foreign atoms which leads to a lowering of energy – see **Fig.10** (chlorine atoms are chequered).

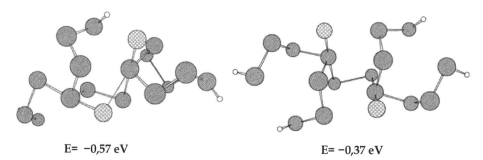

E= −0,57 eV   E= −0,37 eV

Fig. 10. Two of the lowest energy complex HVBs in Se-Cl after [63].

The configurations and energies in the Se-Cl case differ from those of Se-Te (Fig.9); in this way, the specificity of foreign atoms is revealed. Note that in the left configuration of the lowest energy in **Fig.10**, the Cl_2Se_5 fragment has not only over-coordinated Cl (a two-coordinated instead of the usual one-coordinated state) but also five-coordinated Se, i.e. chlorine induces the most coordinated Se state obtained by us so far (see four-coordinated Se in Fig.4, Fig.9 and Fig.10, as well as the configurations with three-coordinated Se in other chalcogenide glasses [39]). One should note, however, that the relation between coordination and valence is not strictly determined, especially in the hypervalent case [64]. Indeed, by use of only the geometric arrangement, it is impossible to conclude does a chemical bonding exist between the atoms under consideration – this is the question to the quantum-chemistry study of a concrete configuration.

The method of adding foreign atoms in a simple non-crystalline matrix with the further comparison of experimental features with quantum-chemical models seems to be a fruitful way for understanding what is the nature of chemical bonding in glass formers and how one can creates various bonding states and/or structures. As to the structures, let us consider an additional feature that we found when investigating Se-X glasses by means of IR spectroscopy.

In **Fig.11** one can see a *narrow* absorption band at 792 cm^{-1} on the high-frequency side of the third overtone for the 245 cm^{-1} band of Se: the line half width is one-third of that for the overtone. Note that the 792 cm^{-1} band develops only for a definite concentration of chlorine, existing in the sample 0.01%Cl, which is most persistent for nucleation (Fig.7), and in the sample 0.05%Cl, which has the greatest capacity for heterogeneous nucleation (Fig.8). Interestingly, the same 792 cm^{-1} band also observed in the Se-As series for the 0.25%As sample (compared with the minimal heterogeneity in Fig.8) and unexpectedly at 5%As [65] (one can also see the extrema at the same concentrations of 0.01-0.05%Cl and 0.025%As in other properties for fresh glasses in Fig.5). Hence, the *ordering* which reveals itself as the 792 cm^{-1} narrow band can be related to the selenium matrix, the structure of which is modulated by means of the impurity introduced into it. Interestingly, the ordering can manifest in properties in the opposite manner, depending upon the impurity, as it is seen in Fig.5 for the extrema at 0.01-0.05%Cl and 0.025%As .

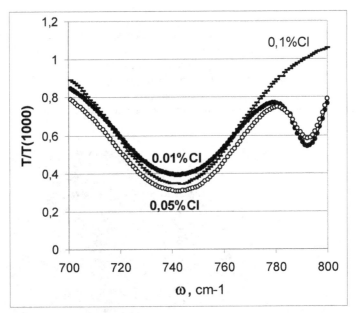

Fig. 11. IR spectra of Se-Cl glasses within the "phonon" range (400-1000 cm^{-1}) after [60]. *T* is the transmission at a given frequency, $T(1000)$ is the transmission at 1000 cm^{-1}, i.e., the *transparency* (see Fig.7 and Fig.8).

When analysing the properties, a thoughtful reader can find out the disagreement in the $r(0)$ values in Fig.5 and in the T_2/T_1 values in Fig.7 for pure selenium glass (X=0) in different Se-X systems, a fact that may devaluate the data itself. The explanation is rather simple: different series were formed in different regimes: the first regime was applied for Se-Te and Se-Ge, the second one for Se-As and Se-S, and the third one for Se-Cl. Indeed, Se glasses have similar properties in the series of the same preparation. This is the *memory effect*, which is well known for anyone who works with glasses.

The non-linearity, spontaneous ordering and the memory effect discussed above are signs of a *self-organizing* system, a system that possesses various scenarios for evolution, depending upon the system's nature and the information provided by external medium [66]. When glass is considered as a self-organizing system with bond waves as the basis for self-organization, one acquires an appropriate tool not only for the explanation of the above-described phenomena but also for the planning of special experiments. In the following experiments, we use an external *information field* for governing the development of a glass structure by means of the directed spreading of bond waves in the region where they are most effective, i.e., at $T>T_g$ where 3D bond waves exist. Note that the above-described experiments concern nucleation in solid glass ($T>T_g$), a process that is provided by 2D bond waves spreading in the limits of frozen wavefronts. Thus, let us defreeze them.

4. The crystallization of softening glass in an ultrasonic field

4.1 Experimental conditions

The same Se-X series of glasses is used and optical transmission is also used as the general method for the observation of the crystallization process. The distinctions from the above experiments are as follows. First, the *temperature* range of the treatment 50-72°C is somewhat *above* the glass transition temperature T_g (which is about 35°C for Se), although the samples retain their form (in general). Second, the *time* of the treatment was 5-10 min instead of 5 years, as in the previous case. Third, the *ultrasonic field* is applied during treatment. Finally, the optical transmission was measured in *two directions* perpendicular to one another, before and after treatment.

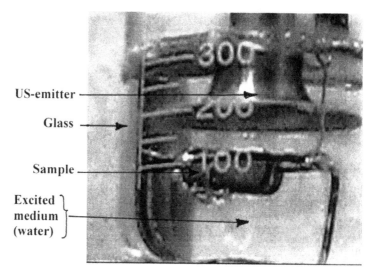

Fig. 12. The treatment of a Se-X sample in an ultrasonic field in a cavitation regime. Note the glass with water, the sample in the holder near the 100 ml mark and the end of a US-emitter at the 200 ml mark, as well as the large bubbles at the water's surface and near the sample.

In the framework of our understanding of glass as a self-organizing system, owing to the collective behaviour of hypervalent bonds in a covalent network, these conditions mean, first, the existence of 3D bond waves which activate *all of the volume* of a substance. Second, the crystallization process *accelerates* owing to not only the increase of the bond wave dimensionality but also of the bond wave velocity. Third, an ultrasonic field can play the role of an *information field* that can orient bond waves. Finally, *anisotropy* can arise after the treatment.

The experimental equipment shown in **Fig.12** is rather simple: a glass with water (the excited medium for ultrasonic cavitation), in which the sample is placed in the holder. The sample has two pairs of perpendicular grains - A-A and B-B - the distance between like grains being equal: $d_{A-A}=d_{B-B}$. The US-emitter (its own frequency is 24 kHz and its vibration amplitude is 3 μ) is disposed at 10 mm above the upper grain A. The temperature varied from 50ºC to 72ºC and the time of treatment from 10 to 5 min. Note that the field is *weak*, having an intensity of about 0.2-0.3 W/cm² (the cavitation threshold is 0.1 W/cm²). The vibration frequency in the excited medium is within the range of 1 kHz - 1 GHz, with a maximum at 5-10 MHz. At the end of treatment, the US-input is turned off, the emitter is lifted and the sample together with the holder is taken out of the glass with further cooling in the air down to room temperature. Then all the samples of a given Se-X series are subjected to optical investigation, which is performed at the same day.

The optical measurements were made in two directions: in the A-A direction along the axis of the US-emitter (i.e., through two "frontal" A grains, which are parallel to the emitter end in Fig.12) and in the B-B direction perpendicular to the emitter axis (one can see one of the lateral grains B facets in face in Fig.12). These directions are easily distinguished in the samples owing to the curve sides remaining after transformation of initial cylinders/tablets, which were used in the previous experiments at $T<T_g$, into the blocks using in the present experiments; the A-A grains correspond to the ends of the initial cylinders. The spectra were measured within the 400-5000 cm⁻¹ range and the data up to 1000 cm⁻¹ are discussed here. It may be desirable to catch the eye of future investigators to the 1000-5000 cm⁻¹ interval: although we do not discuss the related data, it is a quite sensitive area for the observation of the crystallization process in the scale from 10μ to 2μ, respectively.

4.2 Basic experimental features

This study is currently in progress, although two articles describing experiments on the Se-As series [65] and Se-Te series [67] have just become available in English. Here, only basic experimental features are discussed; for additional information the reader is invited to look at the originals. Note that we use the optical transmission method for watching the crystallization process, which is again attributed primarily to the nucleation stage, a process that decreases transmission due to light scattering from a well-developed internal glass/crystal surface.

The first feature is a very fast darkening (for minutes) as compared with the previously considered darkening due to ageing (years). Thus, the first question is: what causes such rapid darkening? Temperature or cavitation?

It can be seen in **Fig.13** that even at 72ºC - the highest temperature which was applied - the temperature itself has only a low effectiveness (compare the spectra 12 to 13) and only the US treatment strongly accelerates nucleation in softening glass (see spectrum 5 of a very low intensity).

The second feature is optical *anisotropy*, which is very weak in the initial glasses but can develop after US treatment. The anisotropy value in a given series depends on composition, as can be seen in **Fig.14** with Se-As as an example. Despite the series investigated, anisotropy always arises as a result of a strong darkening in the A-A direction together with a negligible change in transparency in the B-B direction. As to the anisotropy peak position, a special concentration of 0.25%As in Fig.14 corresponds with the above-mentioned appearance of a narrow 792 cm^{-1} band - the same as that which is shown in Fig.11 for the Se-Cl series. This fact establishes a connection between the ability of pre-crystalline ordering in an as-prepared glass and the ability for the development of gigantic optical anisotropy in the glass after its further high-temperature treatment in an ultrasonic field.

In the framework of the bond wave model, anisotropy arises due to the definite location of an excitation source (see Fig.12): although one can suppose a nearly spherical distribution for the dissipation of the input energy in a cavitating medium, the input itself violates the symmetry. The excited medium, in addition, has a temperature above the glass transition temperature for the Se-X glasses investigated. Therefore, when an initially isotropic glass is placed into the cell (Fig.12), the bond waves - which were initially frozen in different directions, as in the case shown in the right-hand part of Fig.3 - begin to refreeze (2D→3D), and the refrozen bond wave will move in a definite direction, remaining this direction after glass cooling.

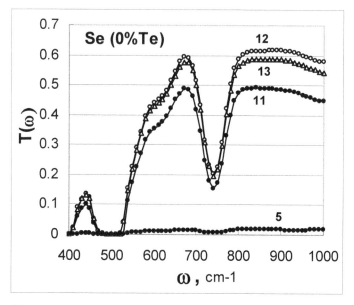

Fig. 13. Optical transmission spectra of the aged Se glass (11), the aged glass after polishing (12) and the polished glass after a 5 min treatment at 72°C (13). Curve 5 corresponds with the polished sample (an analogue of 12) after a 5 min treatment in a cavitation US field at the same temperature, 72°C. The measurements are made in the A-A direction.

The question is: in what direction will the refrozen bond waves move? It seems likely that the layers/wavefronts will be oriented in the A-A direction, parallel to the vibrating end of the US-emitter (see Fig.13). Insofar as crystallization begins within the layers containing active hypervalent bonds, then one obtains a system of partially-crystallized layers divided by periodic glass-like regions. This system is frozen in the sample after its removal from the cell with further quenching in the air. The sample reveals optical anisotropy (see line 5 in Fig.14) because in such a layered structure a measuring beam should scatter strongly when it probes in the A-A direction (the beam falls onto the nucleated layers) and scatter much less in the B-B direction (the beam spreads within transparent glass-like regions between the layers, as like along the optical wire).

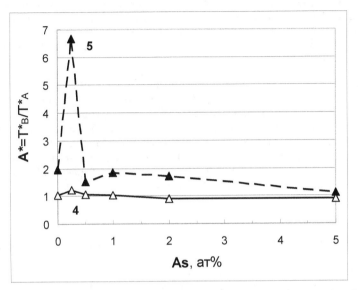

Fig. 14. The anisotropy in transparency (transmission at 1000 cm^{-1} measured in the A-A and B-B directions) for the Se-As series. The two lines correspond with the measurements before (4) and after (5) US treatment (72^{0}C, 5 min).

In an attempt to see the proposed internal structure, we used electron microscopy; the two images shown in **Fig.15** correspond to fresh fractures made in the two perpendicular directions.

Fig. 15. A SEM image of the fracture made parallel to the A grain (on the left) and the B grain (on the right) for a selenium sample after US treatment at 72°C.

In order to exclude the effects of many-component crystallization, we made the fractures using pure Se glass subjected to US treatment. This is that glass belonging to the Se-As series with the induced anisotropy of A*=2 (see 0% in Fig.14): after the US treatment, the transparency in the A-A direction was decreased doubly, while the transparency in the B-B direction was unchanged. In **Fig.15**, we present two images for the two fractures. The left image probably corresponds to the crystallized d-layer, along which a fracture developed. The right image looks like ordinary glass, probably corresponding to the glass-like region between the two adjacent d-layers.

For the crystallized layer (Fig.15, on the left) it is interesting to note the unusual *needle* form of nucleation in the selenium, which is known to be inclined to form the spheric-like crystallites. Of course, a much more intensive SEM investigation, as well as a wide set of other methods, is needed for a more adequate characterization of the materials obtained and, consequently, a deeper understanding of the processes involved in their formation. We hope that the described experiments of crystallization in glass under the action of impurities and/or ultrasonic field will help the coming investigators of glass and glass-ceramics in obtaining the new materials using non-traditional ways.

5. Conclusions

The structure of glass, considered from the chemical bond point of view using the bond wave model, acquires specific elements of order which develop during cooling (open system) owing to the alternation of chemical bonds (two-stable element) in the form of a bond wave (collective feedback between the elements). In the brackets, there are three general conditions for self-organization; they are naturally fulfilled in glass forming substances owing to their ability to form alternative hypervalent bonds. The processes in a self-organizing system proceed in a specific manner, and this is the reason why classic notions about crystallization works poorly in glass formers. On the other hand, self-organization opens new possibilities for both the understanding and management of related materials. One should note that although "self-organization in glass" becomes a hot topic (see [68] as an example), a real chemical bond is actually absent in the related works. We hope that our approach, connecting contemporary notions about chemical bonding - from the one side - and the self-organization theory (synergetics) - from the other side - will help to fill a significant gap in glass theory, including the understanding of the mechanism of glass-crystal transition and, in glass practice, the elaboration of new principles for the formation of glass-ceramic materials.

6. Acknowledgements

Dedicated to Prof. S. A. Dembovsky (1932-2010); his ideas and our long-term collaboration actually make him the co-author of this work.

This work was supported by the Russian Foundation of Basic Research (Grant No. 09-03-01158).

7. References

[1] J. Zarzycki. Glasses and the vitreous state. Cambridge University Press, Cambridge-New-York-Melbourne, 1991.

[2] P.T. Sarjeant, R. Roy. Mater. Res. Bull., 3, 265 (1968).
[3] D.R. Uhlmann. J. Non-Cryst. Solids, 7, 337 (1972).
[4] J.C.A. Vreeswijk, R.G. Gossink, J.M. Stevels. J. Non-Cryst. Solids, 16, 15 (1974).
[5] E. Ruckenstein, S.K. Ihm. J. Chem. Soc. Faraday Trans. I, 72, 764 (1976).
[6] G.F. Neilson, M.C. Weinberg. J. Non-Cryst. Solids, 34, 137 (1979).
[7] R.J.H. Gelsing, H.N. Stein, J.M. Stevels. Phys. Chem. Glasses, 7, 185 (1966).
[8] J.C.Th.G.M. Van der Wielen, H.N. Stein, J.M. Stevels. J. Non-Cryst. Solids, 1, 18 (1968).
[9] A.C.J. Havermans, H.N. Stein, J.M. Stevels. J. Non-Cryst. Solids, 5, 66 (1970).
[10] M.H.C. Baeten, H.N. Stein, J.M. Stevels. Silicates Industriels, 37, 33 (1972).
[11] G. Whichard, D.E. Day. J. Non-Cryst. Solids, 66, 477 (1984).
[12] W. Huang, C.S. Ray, D.E. Day. J. Non-Cryst. Solids, 86, 204 (1986).
[13] M.D. Mikhailov, A.C. Tver'janovich. Glass Phys. Chem., 6, No.5 (1980).
[14] M.D. Mikhailov, A.C. Tver'janovich. Glass Phys. Chem., 12, No.3 (1986).
[15] S.A. Dembovsky, L.M. Ilizarov, E.A. Chechetkina, A.Yu. Khar'kovsky. In: Proc. "Amorphous Semiconductors – 84". Gabrovo, Bulgaria, 1984. p.39.
[16] S.A. Dembovsky, L.M. Ilizarov, A.Yu. Khar'kovsky. Mater. Res. Bull. 21, 1277 (1986).
[17] S.A. Dembovsky, E.A. Chechetkina. Glass Formation (in Russ.). Nauka, Moscow, 1990.
[18] K.F. Kelton, A.L. Greer. J. Non-Cryst. Solids, 79, 295 (1986).
[19] S.A. Dembovsky. Zh. Neorg. Khim. (Russ. J. Inorg. Chem.), 22, 3187 (1977).
[20] S.A. Dembovsky. Proc. "Amorphous Semiconductors – 80". Kishinev, USSR, 1980. p.22.
[21] S.A. Dembovsky, E.A. Chechetkina. Mater. Res. Bull., 16, 606 (1981).
[22] S.A. Dembovsky, E.A. Chechetkina. Mater. Res. Bull., 16, 723 (1981).
[23] S.A. Dembovsky, E.A. Chechetkina. J. Non-Cryst. Solids 64, 95 (1984).
[24] V.M. Fokin, E.D. Zanotto, W.P. Schmelzer, O.V. Potapov. J. Non-Cryst. Solids, 351, 1491 (2005).
[25] J.L. Nowinski, M. Mroczkowska, J.E. Garbarczyk, M. Wasiucionek. Materials Science-Poland 24, 161 (2006).
[26] A.H. Moharram, A.A. Abu-sehly et al. Physica B, 324, 344 (2002).
[27] L. Berthier. Physics 4, 42 (2011).
[28] P.N. Pusley, E. Zaccarelli et al. Phil. Trans. R. Soc. A, 367, 4993 (2009).
[29] H. Tanaka. J. Phys.: Condens. Matter, 23, 284115 (2011).
[30] W.H. Zachariasen. J. Amer. Ceram. Soc., 54, 3841(1932).
[31] G. Biroli. Sèminaire Poincarè XII, 37 (2009).
[32] S.A. Dembovsky. Mater. Res. Bull., 16, 1331 (1981).
[33] N.A. Popov. JETP Lett., 31, 409 (1980).
[34] S.A. Dembovsky, E.A. Chechetkina. J. Non-Cryst. Solids, 85, 346 (1986).
[35] S.A. Dembovsky, E.A. Chechetkina. Philos. Mag., B53, 367 (1986).
[36] S.A. Dembovsky, E.A. Chechetkina. J. Optoel. Adv. Mater., 3, 3 (2001).
[37] S.A. Dembovsky. J. Non-Cryst. Solids, 353, 2944 (2007).
[38] E.A. Chechetkina. J. Optoel. Adv. Mater., 13, 1385 (2011).
[39] S.A. Dembovsky, A.S. Zyubin. Russ. J. Inorg. Chem., 46, 121 (2001).
[40] S.A. Dembovsky, A.S. Zyubin, O.A. Kondakova. Mendeleev Chemistry Journal (Zhurnal Ross. Khim. Ob-va im.D.I.Mendeleeva), 45, 92 (2001).
[41] S. Zyubin, F.V. Grigoriev, S.A. Dembovsky. Russ. J. Inorg. Chem., 46, 1350 (2001).
[42] J.L. Muscher. Angew. Chem. Int. Ed., 8, 54 (1969).
[43] W.B. Jensen. J. Chem. Educ., 83, 1751 (2006).

[44] G. Pimentel. J. Chem. Phys., 19, 446 (1981).

[45] S.C. Moss, D.L. Price. In: Physics of Disordered Materials (Eds. D. Adler, H. Fritzsche, S.R. Ovshinsky), Plenum, New York, 1985. p.77.

[46] E.A. Chechetkina. J. Phys.: Condes. Matter, 7, 3099 (1995).

[47] E.A. Chechetkina. Solid State Commun., 87, 171 (1993).

[48] E.A. Chechetkina. In: Proc XVII Inter. Congr. Glass, Beijing, 1995. vol.2, p.285.

[49] E.A. Chechetkina. J. Non-Cryst. Solids, 128, 30 (1991).

[50] E.A. Chechetkina. In: Fractal Concepts in Materials Science. MRS Proc., vol.367 (1995).

[51] E.A. Chechetkina. J. Non-Cryst. Solids, 201, 146 (1996).

[52] S.A. Dembovsky, A.S. Zyubin. Russ. J. Inorg. Chem. 54, 455 (2009).

[53] M. Kastner, D. Adler, H. Fritzsche. Phys. Rev. Lett. 37, 1504 (1976).

[54] S.A. Dembovsky, E.A. Chechetkina, T.A. Kupriyanova. Materialovedenie (Mater. Sci. Trans.), No.4, 37 (2004).

[55] M.F. Kotkata, E.A. Mahmoud, M.K. El-Mously. Acta Phys. Acad. Sci. Hung., 50, 61 (1981).

[56] M.F. Kotkata, M.K. El-Mously. Acta Phys. Hung., 54, 303 (1983).

[57] G. Parthasarathy, K.J. Rao, E.S.R. Gopal. Philos. Mag., B50, 335 (1984).

[58] S.A. Dembovsky, L.M. Ilizarov, A.Yu. Khar'kovsky. Mater. Res. Bull., 21, 1277 (1986).

[59] M. Hansen, K. Anderko. Constitution of Binary Alloys. McGraw-Hill, 1958.

[60] E.A. Chechetkina, A.B. Vargunin, V.A. Kuznetzov, S.A. Dembovsky. Materialovedenie (Mater. Sci. Trans.), No.7, 28 (2006).

[61] E.A. Chechetkina, E.B. Kryukova, A.B. Vargunin. Materialovedenie (Mater. Sci. Trans.), No.6, 29 (2007).

[62] A.S. Zyubin, E.A. Chechetkina, S.A. Dembovsky. Russ. J. Inorg. Chem., 57, 913 (2012).

[63] S. Zyubin, S.A. Dembovsky. Russ. J. Inorg. Chem., 56, 329 (2011).

[64] D.W. Smith. J. Chem. Educ., 82, 1202 (2005).

[65] E.A. Chechetkina, E.V. Kisterev, E.B. Kryukova, A.I. Vargunin. Inorg. Mater.: Appl. Res., 2, 360 (2011).

[66] H. Haken. Information and Self-Organization. A macroscopic Approach to Complex Systems. Springer, 1988, 2000, 2006.

[67] E.A. Chechetkina, E.V. Kisterev, E.B. Kryukova, A.I. Vargunin. J. Optoel. Adv. Mater., 11, 2034 (2009).

[68] P. Boolchand, G. Lucovsky, J.C. Phillips, M.F. Thorpe. Philos. Mag., 85, 3823 (2005).

Numerical Models of Crystallization and Its Direction for Metal and Ceramic Materials in Technical Application

Frantisek Kavicka[1], Karel Stransky[1], Jana Dobrovska[2],
Bohumil Sekanina[1], Josef Stetina[1] and Jaromir Heger[1]
[1]Brno University of Technology,
[2]VSB TU Ostrava
Czech Republic

1. Introduction

Structure of metallic and also majority of ceramic alloys is one of the factors, which significantly influence their physical and mechanical properties. Formation of structure is strongly affected by production technology, casting and solidification of these alloys. Solidification is a critical factor in the materials industry, e.g. (Chvorinov, 1954). Solute segregation either on the macro- or micro-scale is sometimes the cause of unacceptable products due to poor mechanical properties of the resulting non-equilibrium phases. In the areas of more important solute segregation there occurs weakening of bonds between atoms and mechanical properties of material degrade. Heterogeneity of distribution of components is a function of solubility in solid and liquid phases. During solidification a solute can concentrate in inter-dendritic areas above the value of its maximum solubility in solid phase. Solute diffusion in solid phase is a limiting factor for this process, since diffusion coefficient in solid phase is lower by three up to five orders than in the melt (Smrha, 1983). When analysing solidification of these alloys so far no unified theoretical model was created, which would describe this complex heterogeneous process as a whole. During the last fifty years many approaches with more or less limiting assumptions were developed. Solidification models and simulations have been carried out for both macroscopic and microscopic scales. The most elaborate numerical models can predict micro-segregation with comparatively high precision. The main limiting factor of all existing mathematical micro-segregation models consists in lack of available thermodynamic and kinetic data, especially for systems of higher orders. There is also little experimental data to check the models (Kraft & Chang, 1997).

Many authors deal with issues related to modelling of a non-equilibrium crystallisation of alloys. However, majority of the presented works concentrates mainly on investigation of modelling of micro-segregation of binary alloys, or on segregation of elements for special cases of crystallisation – directional solidification, zonal melting, one-dimensional thermal field, etc. Moreover these models work with highly limiting assumption concerning phase diagrams (constant distribution coefficients) and development of dendritic morphology (mostly one-dimensional models of dendrites); e.g. overview works (Boettingen, 2000; Rappaz, 1989; Stefanescu, 1995). Comprehensive studies of solidification for higher order real alloys are rarer. Nevertheless, there is a strong industrial need to investigate and simulate more complex alloys because nearly all current commercial alloys have many components often exceeding ten elements. Moreover, computer simulation have shown that even minute amounts of alloying elements can significantly influence microstructure and micro-segregation and cannot be neglected (Kraft & Chang, 1997). Dendritic crystallisation is general form of crystallisation of salts, metals and alloys. At crystallisation of salts from solutions a dendritic growth of crystals occurs at high crystallisation rate, which requires high degree of over-saturation. Findings acquired at investigation of crystallisation of salts were confirmed at investigation of crystallisation of metals. If negative temperature gradient is present in the melt before the solidification front, this leads to a disintegration of the crystallisation front and to formation of dendritic crystallisation (Davies, 1973). High crystallisation rate is characteristic feature of dendritic crystallisation. Solutes have principal influence on the crystallisation character, as they are the cause of melt supercooling before the crystallisation front and formation of the negative temperature gradient. This kind of supercooling is called constitutional supercooling. For example a layer of supercooled melt is formed in a steel ingot in the immediate vicinity of the interface melt-solid, in principle at the very beginning of crystallisation as a result of segregation of solutes, which causes decrease in solidification temperature of this enriched steel. Increased concentration of solutes creates soon a broad zone of constitutionally supercooled steel, in which the crystallisation rate is high. During subsequent solidification, when the crystallisation rate is low, the value of temperature gradient is also low, which means that conditions for dendritic crystallisation are fulfilled again (Šmrha, 1983). More detailed information on dendritic crystallisation – see classical works (Chalmers, 1940), (Flemings, 1973), (Kurz &Fisher, 1986). According Chvorinov (1954), Smrha (1983) and others metallic alloys and also majority of ceramic alloys in technical application are always characterised by their dendritic crystallisation. It is therefore of utmost importance that their final desirable dendritic structure has appropriate properties that can be used in technical practice. These properties depend of the kind of practical use and they comprise flexibility, elasticity, tensile strength, hardness, but also for example toughness. In the case of ceramic materials the properties of importance are brittleness, fragility and very often also refractoriness or resistance to wear. This chapter presents numerical models of crystallisation of steel, ductile cast iron and ceramics EUCOR aimed at optimisation of their production and properties after casting.

2. Numerical models of the temperature field and heterogeneity

Crystallization and dendritic segregation of constitutive elements and admixtures in solidifying (crystallising) and cooling gravitationally cast casting or continuously cast blank (shortly concast blank) is directly dependent on character of formation of its temperature field. Especially rate and duration of the running crystallisation at any place of the blank, so called local solidification time, is important. Solidification and cooling-down of a gravitationally cast casting as well as the simultaneous heating of a metal or non-metal mould is a rather complex problem of transient heat and mass transfer. This process in a system casting- mould-ambient can be described by the Fourier's equation (1) of 3D transient conduction of heat.

$$\frac{\partial T}{\partial \tau} = \frac{\lambda}{\rho c}\left(\frac{\partial^2 T}{\partial x^2} + \frac{\partial^2 T}{\partial y^2} + \frac{\partial^2 T}{\partial z^2}\right) + \frac{Q_{SOURCE}}{\rho c} \tag{1}$$

In equation (1) are T temperature [K], τ time [s], λ heat conductivity [W.m^{-1}.K^{-1}], ρ density [kg.m^{-3}], c specific heat capacity [J.kg^{-1}.K^{-1}], Q_{SOURCE} latent heat of the phase or structural change [W.m^{-3}], x,y,z axes in given directions [m].

Implementation of continuously casting (shortly concasting), which considerably increased rate of melt cooling between temperatures of liquidus and solidus brought about time necessary for crystallic structure homogenisation. 3D transient temperature field of the of the system of concast blank-mould or concast blank-ambient is described by Fourier-Kirchoff's equation (2).

$$\frac{\partial T}{\partial \tau} = \frac{\lambda}{\rho c}\left(\frac{\partial^2 T}{\partial x^2} + \frac{\partial^2 T}{\partial y^2} + \frac{\partial^2 T}{\partial z^2}\right) + \left(w_x\frac{\partial T}{\partial x} + w_y\frac{\partial T}{\partial y} + w_z\frac{\partial T}{\partial z}\right) + \frac{Q_{SOURCE}}{\rho c} \tag{2}$$

In equation (2) w_x, w_y, w_z are velocity in given directions [m.s^{-1}]

These equations are solvable only by means of modern numerical methods. Therefore original models of the transient temperature field (models **A**) of both systems of gravitational casting or continuous casting were developed. Both models are based on the 1st and 2nd Fourier's laws on transient heat conduction, and the 1st and 2nd law of thermodynamics. They are based on the numerical method of finite differences with explicit formula for the unknown temperature of the mesh node in the next time step, which is a function of temperatures of the same node and six adjacent nodes in Cartesian coordinate system in the previous time step. Models take into account non-linearity of the task, it means dependence of thermo-physical properties of all materials of the systems on temperature and dependence of heat transfer coefficients on temperature of all external surfaces. Models are equipped with and interactive graphical environment for automatic generation of a mesh, and for evaluation of results, it means by so called pre-processing and post-processing.

Another model, which has also been already mastered is model of chemical heterogeneity of chemical elements (model **B**), enables description and measurement of dendritic segregation of constitutive elements and admixtures in crystallising and cooling blank (casting or concasting). This model is based on the 1st and 2nd Fick's laws of diffusion and it comprises implicitly also the law of conservation of mass. The solution itself is based on the Nernst distribution law, which quantifies at crystallisation distribution of chemical elements between liquid and solid phases of currently crystallising material in the so called mushy zone (i.e. in the zone lying between the temperature of liquidus and solidus). Majority of parameters necessary for application of the models **A** is known, but parameters necessary for use of the model **B** had to be determined by measurements on the work itself, i.e. on suitably chosen samples from continuously crystallised blanks.

Measurement was realised in the following manner: at selected segments of the cast blank concentration of main constitutive, additive and admixture elements was determined from the samples taken in regular steps. In dependence on chemical heterogeneity and structure of a blank the segments with length of 500 to 1000 μm were selected, and total number of steps, in which concentration of elements was measured, was set to 101. Measurement of concentration of elements was performed by methods of quantitative energy dispersive analysis (EDA) and wave dispersive analysis (WDA) of X-ray spectral microanalysis, for which special software and special measurement device was developed for use in combination with the analytical complex JEOL JXA 8600/KEVEX.

After completion of measurement the sample surface was etched in order to make visible the original contamination of surface by electron beam, and the measured traces were photographically documented, including the mean distance of dendrites axes within the measured segment. It was verified that the basic set of measured concentration data of elements (8 to 11 elements) makes it possible to obtain a semi-quantitative to quantitative information on chemical heterogeneity of the blank, and that it is possible to apply at the same time for evaluation of distribution of elements in the blank structure the methods of mathematical- statistical analysis. It is possible to determine the distribution curve of the element concentration in the measured segment of the analysed blank and their effective distribution coefficient between the solid and liquid phase during crystallisation. In this way the crucial verified data necessary for creation of the conjugated model (**AB**) of crystallising, solidifying and cooling down blank were obtained. It was verified that re-distribution of constitutive, additive elements and admixtures can be described by effective distribution coefficient, which had been derived for parabolic growth by Brody and Flemings (Brody&Flemings, 1966). At the moment of completed crystallisation, at surpassing of an isosolidic curve in the blank, it is possible to express the ratio of concentration of dendritically segregating element C_S to the mean concentration of the same element at the given point of the blank C_0 by the relation

$$C_S/C_0 = k_{ef}[1 - (1 - 2\alpha k_{ef})g_S]^{(k_{ef}-1)/(1-2\alpha)} \tag{3}$$

where k_{ef} is effective distribution coefficient, g_S is mass share of the solidified phase, and α is dimensionless Fourier's number of the 2nd kind for mass transfer. This number is given by the relation

$$\alpha = D_S \theta_{lS}/L^2 \tag{4}$$

in which D_S is diffusion coefficient of the segregating element in solid phase, θ_{lS} is local time of crystallisation (i.e. time of persistence of the assumed dendrite between the temperature of liquidus and solidus) and L is mean half distance of dendritic axes (namely of axes of secondary dendrites). In the next step it is necessary to express the ratio of concentrations C_S/C_0 express as a function of concrete heterogeneity index I_H and of statistical distribution of the measured element, expressed by distributive curve of crystallization segregation. In this manner the following equation is available for each measured element:

$$C_S/C_0 = I_H \tag{5}$$

which expresses by concrete numbers the parameters, defined by the equation (3). By solving equations (3, 4) it is then possible to determine for each analysed element (i.e. for its measured index of dendritic heterogenity, effective distribution coefficient and distribution curve of dendritic segregation, i.e. for the established statistic character of distribution of the analysed element in structure of the blank) certain values of dimensionless criterion α. Afterwards on the basis of semi-empiric relations and rates of movement of the crystallisation front, calculated from the thermal field model in confrontation with the results of experimental metallographic analysis, the mean value of distances between branches of secondary dendrites L was determined for 9 samples of the blank. The values θ_{lS} and L for the criterion α, are calculated from the model for each sample, which were determined for individual measured elements in each sample of the blank. It is possible to make from the equation (4) an estimate of the diffusion coefficient of each analysed element in individual samples of the blank. At the moment, when temperature of any point of the mesh drops below the liquidus temperature, it is valid that the share of the forming solid phase g_s grows till its limit value $g_s = 1$ (i.e. in solid phase). In this case segregation of the investigated element achieves in the residual inter-dendritic melt its maximum.

The combination of mentioned models and methodology of chemical heterogeneity investigation are presented on following technical applications.

3. Gravitational casting

3.1 Solidification of massive casting of ductile cast iron

The quality of a massive casting of cast iron with spheroidal graphite is determined by all the parameters and factors that affect the metallographic process and also others. This means the factors from sorting, melting in, modification and inoculation, casting, solidification and cooling inside the mould and heat treatment. The centre of focus were not only the purely practical questions relating to metallurgy and foundry technology, but mainly the verification of the possibility of applying two original models – the 3D model of transient solidification and the cooling of a massive cast-iron casting and the model of chemical and structural heterogeneity.

3.1.1 Calculation and meassurement of the temperature field

The application of the 3D numerical model on a transient temperature field requires systematic experimentation, including the relevant measurement of the operational parameters directly in the foundry. A real 500×1000×500 mm ductile cast-iron block had been used for the numerical calculation and the experiment. They were cast into sand moulds with various arrangements of steel chills of cylindrical shape. The dimensions of the selectid casting, the mould, the chills and their arrangements are illustrated in Figure 1. The

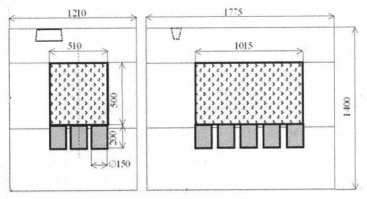

Fig. 1. The forming of casting no. 1 with chills on one side

courses of the temperatures on casting No. 1 were measured for 19 hours 11 min after pouring. The iso-zones, calculated in castings No.1 and in the chills in the total solidification time after casting, are illustrated in Figure 2 (Dobrovska et al., 2010).

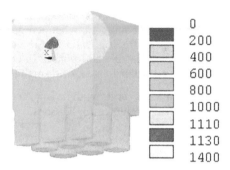

Fig. 2. The calculated iso-zones in casting No. 1 (5 hours)

3.1.2 The relation between the model of the temperature field and the model of structural and chemical heterogeneity

The 3D numerical model of the temperature field of a system comprising the casting-mold-ambient is based on the numerical finite-element method. The software ANSYS had been chosen for this computation because it enables the application of the most convenient method of numerical simulation of the release of latent heats of phase and structural changes using the thermodynamic enthalpy function. The software also considers the non-linearity of the task, i.e. the dependence of the thermophysical properties (of all materials entering the system) on the temperature, and the dependence of the heat-transfer coefficients (on all boundaries of the system) on the temperature of the surface — of the casting and mold. The original numerical model had been developed and used for estimation of structural and chemical heterogeneity. Initial and boundary conditions had been defined by means of theory of similarity. The verifying numerical calculation of the local solidification times θ – conducted according to the 3D model proved that, along the height, width and length of these massive castings, there are various points with differences in the solidification times of up to two orders. The aim was to verify the extent to which the revealed differences in the local solidification times affect the following parameters (Stransky et al., 2010):

a.   The average size of the spheroidal graphite particles;
b.   The average density of the spheroidal graphite particles;
c.   The average dimensions of the graphite cells, and
d.   The chemical heterogeneity of the elements in the cross-sections of individual graphite cells.

The relationships – among the given four parameters and the corresponding local solidification times – were determined in the series of samples that had been selected from defined positions of the massive casting. The bottom part of its sand mould was lined with (a total number of) 15 cylindrical chills of a diameter of 150 mm and a height of 200 mm. The upper part of the mould was not lined with any chills. The average chemical composition of the cast-iron before casting is given in Table 1.

Element	C	Mn	Si	P	S	Ti	Al	Cr	Ni	Mg
wt.%	3.75	0.12	2.15	0.039	0.004	0.01	0.013	0.07	0.03	0.045

Table 1. Chemical composition of ductile cast-iron (casting No. 1)

A 500×500×40 mm plate had been mechanically cut out of the middle of the length by two parallel transversal cuts. Then, further samples were taken from exactly defined points and tested in terms of their structural parameters and chemical heterogeneity. Samples in the form of testing test-samples for ductility testing, with threaded ends, were taken from the bottom part of the casting (A), from the middle part (C) and from the upper part (G). The 15 mm in diameter and 12 mm high cylindrical samples served the actual measurements in order to determine the structural parameters and chemical heterogeneity. In the points of the defined positions of the samples prepared in this way, the quantitative metallographic analysis was used to establish the structural parameters of cast-iron, the in-line point analysis to establish the chemical composition of selected elements and numerical calculation using the 3D model to establish the local solidification time. Quantitative analyses of the basic micro-structural parameters in the samples have been the subject of a special study. On each sample a total number of 49 views were evaluated. On the basis of average values of these results the structural parameters of graphite, i.e. the radius of the spheroids of graphite – R_g, the distances between the edges of graphite particles – L_g and the radius of the graphite cells – R_c have been determined for each sample. The concentration of selected elements in each of the samples was measured on the line of L_g between the edges of two particles of spheroidal graphite. The actual measurements of concentrations of ten elements – Mg, Al, Si, P, S, Ti, Cr, Mn, Fe, Ni – was carried out. On each of the samples, the concentrations of all ten elements had been measured in three intervals with each individual step being 3 μm. Before the actual measurement, the regions were selected on an unetched part of the surface and marked with a micro-hardness tester. After the micro-analysis, the samples were etched with nitric acid in alcohol in order to make the contamination of the ground surface visible using an electron beam. Then, using a Neophot light microscope, the interval within which the concentrations were measured was documented. The method of selection of the measurement points is illustrated in Figures 3a,b.

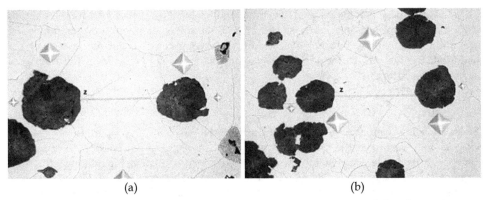

(a)   (b)

Fig. 3. An example of the chemical micro-heterogeneity measurement of ductile cast-iron
a) L_g = 165 μm, b) L_g = 167 μm (in Fig. $L_g \equiv z$)

The results of the measurements of the chemical heterogeneity were evaluated statistically. The element heterogeneity index $I_{H(i)}$ is defined by the quotient of standard deviation of element concentration $\sigma_{c(i)}$ and average element concentration C_i^{av} in the analysed area, i.e. $I_{H(i)} = \sigma_{c(i)}/C_i^{av}$. The element segregation index $I_{S(i)}$ is defined by the quotient of element maximum concentration C_i^{max} and average element concentration C_i^{av} in the analysed area, i.e. $I_{S(i)} = C_i^{max}/C_i^{av}$. The local solidification times of the selected samples of known coordinates within the massive experimental casting were calculated by the 3D model. The calculation of the temperature of the liquidus and solidus for a melt with a composition according to the data in Table 1 was performed using special software with the values: 1130 °C and 1110 °C (the liquidus and solidus temperature). If the local solidification time is known, then it is also possible to determine the average rate of cooling w of the mushy zone as a quotient of the temperature interval and the local solidification time $w = \Delta T/\theta$ [°C/s]. It is obvious from the results that in the vertical direction y from the bottom of the massive casting (sample A: $y = 50$ mm) to the top (gradually samples C: $y = 210$ mm and G: $y = 450$ mm) the characteristic and significant relations are as follows:

a. The average size of the spheroids of graphite R_g, the average size of the cells of graphite R_c and also the average distance between the individual particles of the graphite L_g are all increasing. This relation was confirmed by the quantitative metallographic analysis.
b. The chemical heterogeneity within the individual graphite cells is also changing. The increase in the chemical heterogeneity is reflected most significantly in the increase in the indexes of segregation I_S for titanium which are increasing in the direction from the bottom of the massive casting to the top in the following order: $I_{S(Ti)} = 5.79$-to-9.39-to-11.62
c. The local solidification time increases very significantly – from the bottom of the casting to the top – from 48 s more than 50× (near the centre of the casting) and 95× (at the top of the massive casting).

The relationships between the structural characteristics of graphite in the casting and the local solidification time were expressed quantitatively using a semi-logarithmic dependence:

$$R_g = 19.08 + 2.274\ln\theta \, , R_c = 61.33 + 5.567\ln\theta \, , \ L_g = 84.50 + 6.586\ln\theta \ \ [\mu m, s] \quad (6, 7, 8)$$

As far as chemical heterogenity of the measured elements is concerned, the analogous relation was established only for the dependence of the segregation index of titanium on the local solidification time, which has a steadily increasing course from the bottom of the casting all the way up to the top. The relevant relation was expressed in the form of a logarithmic equation

$$\ln I_S^{Ti} = 1.201 + 0.1410\ln\theta_{\ ls,} \quad [s] \tag{9}$$

The local solidification time θ naturally affects the mechanical properties of cast-iron however with regard to the dimensions of the test pieces; it is not possible to assign the entire body a single local solidification time. To assess relationship among structural parameters, chemical microheterogenity and mechanical properties of analysed cast-iron casting, the selected mechanical properties have been measured. The samples for testing of the tensile strength were taken from the test-sample of the experimental casting in such a way that one had been taken from under the metallographic sample and the second was taken from above. The testing indicates that the local solidification time θ has significant influence on the ductility A_5. The relationship between the ductility and the local solidification time (equation 10) indicates that the reduction in the ductility of cast-iron in the state immediately after pouring is – in the first approximation – directly proportional to the square of the local solidification time.

$$A_5 = 23.399 - 8.1703\theta^2 \qquad [\%, hr] \qquad (10)$$

It can be seen from previous experimentation and the evaluations of the results that – in the general case of the solidification of ductile cast-iron – there could be a dependence of the size of the spheroids of graphite, the size of the graphite cells and therefore even the distance among the graphite particles on the local solidification time. The described connection with the 3D model of a transient temperature field, which makes it possible to determine the local solidification time, seems to be the means via which it is possible to estimate the differences in structural characteristics of graphite in cast-iron and also the effect of the local solidification time on ductility in the poured casting. The main economic goal observed is the saving of liquid material, moulding and insulation materials, the saving of energy and the already mentioned optimization of pouring and the properties of the cast product.

3.2 Casting of corundo-baddeleyit ceramic material

The corundo-baddeleyit material (CBM) belongs to the not too well known system of the Al_2O_3-SiO_2-ZrO_2 oxide ceramics. Throughout the world, it is produced only in several plants, in the Czech Republic under the name of EUCOR. This production process entails solely the utilisation of waste material from relined furnaces from glass-manufacturing plants. EUCOR is heat resistant, wear resistant even at extreme temperatures and it is also resistant to corrosion. It was shown that from the foundry property viewpoint, EUCOR has certain characteristics that are similar to the behavior of cast metal materials, especially steel for castings (Chvorinov, 1954; Smrha, 1983 and others).

3.2.1 Measurements and computation results (the original riser)

The assignment focussed on the investigation of the transient 3D temperature field of a system comprising a casting-and-riser, the mold and ambient, using a numerical model (Heger et al., 2002; Kavicka et. al., 2010) . The dimensions of the casting — the so-called "stone" — were 400 x 350 x 200 mm (Figure 4). The results attained from the numerical analysis of the temperature field of a solidifying casting and the heating of the mold represent only one quadrant of the system in question. Figure 5 shows the 3D temperature field of the casting with the original riser and the mold at two times after pouring. The riser-mold interface is an interesting place for monitoring. Once this point solidifies, the riser can no longer affect the process inside the casting. The initial temperature of the mold was 20°C. The pouring temperature of the melt was 1800°C. That was approximately 300°C higher, when compared with, for example, the steel pouring temperature. The temperature of the liquidus was 1775°C and the solidus 1765°C. The temperature field was symmetrical along the axes, i.e. it was   sufficient for the investigation of the temperature field of a single quadrant only.

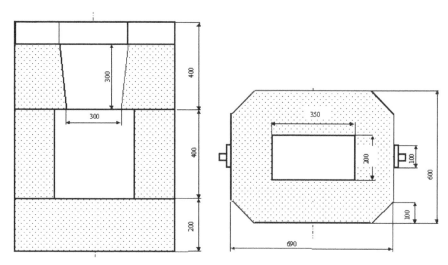

Fig. 4. The casting-riser-mold system

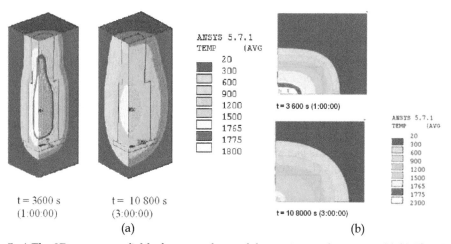

$t = 3600$ s
$(1:00:00)$

$t = 10\,800$ s
$(3:00:00)$

(a)

(b)

Fig. 5. a) The 3D teperature field of one quadrant of the casting wirh riser- mold, b) The 2D temperature field on the riser mold and riser casting interferace

3.2.2 The model of the chemical heterogeneity and its application

The concentration distribution of individual oxides, making up the composition of the ceramic material EUCOR, was determined using an original method (Dobrovska, et al., 2009) and applied in the process of measuring the macro- and micro-heterogeneity of elements within ferrous alloys. This method was initially modified with respect to the differences occurring during solidification of the ceramic material, when compared to ferrous alloys. It was presumed that within EUCOR, the elements had been already distributed, together with oxygen, at the stoichiometric ratio (i.e. the chemical equation), which characterized the resulting composition of the oxides of individual elements after solidification. The preconditions for the application of the model of chemical heterogeneity on the EUCOR material are as follows:

If the analytically expressed distribution of micro-heterogeneity of the oxides of the ceramic material is available, if their effective distribution coefficient is known, and if it is assumed that it is possible to describe the solidification of the ceramic material via analogical models as with the solidification of metal alloys, then it is possible to conduct the experiment on the mutual combination of the calculation of the temperature field of a solidifying ceramic casting with the model describing the chemical heterogeneity of the oxides.

If the Brody-Flemings Model (Brody & Flemings, 1966) is applied for the description of the segregation of oxides of the solidifying ceramic material and if an analogy with metal alloys is assumed, then it is possible to express the relationship between the heterogeneity index I_H of the relevant oxide, its effective distribution coefficient k_{ef} and the dimensionless parameter a using the equation

$$[\ln(2\alpha k_{ef})]/(1 - 2\alpha k_{ef}) = \{\ln[(1 + nI_H^{(m)})/k_{ef}]\}/(k_{ef} - 1) \tag{11}$$

the right-hand side of which $\{\ln[(1+ nI_H^{(m)})/k_{ef}]\}/(k_{ef} - 1)$, based on the measurement of micro-heterogeneity, is already known and through whose solution it is possible to determine the parameter a, which is also on the right-hand side of the equation in $2ak_{ef} = X$. The quantity n has a statistical nature and expresses what percentage of the measured values could be found within the interval $x_s \pm ns_x$ (where x_s is the arithmetic mean and s_x is the standard deviation of the set of values of the measured quantity). If $n = 2$, then 95% of all measured values can be found within this interval. If the dimensionless parameter a is known for each oxide, then a key to the clarification of the relationship exists between the local EUCOR solidification time θ, the diffusion coefficient D of the relevant oxide within the solidifying phase and the structure parameter L, which characterizes the distances between individual dendrites in metalic and ceramic alloys (Figure 6). The equation of the dimensionless parameter a is

$$a = D\theta/L^2 \qquad [\,-\,], [m^2s^{-1}, s, m\,] \tag{12}$$

It is possible to take the dimension of a structure cell as the structure parameter for the EUCOR material. The verification of the possibility of combining both methods was conducted on samples taken from the EUCOR blocks – from the edge (sample B) – and from the centre underneath the riser (sample C). Both the measured and the computed parameters of chemical micro-heterogeneity and the computed parameters of the local solidification time θ (according to the temperature-field model) were calculated. The local solidification time of the sample B was $\theta_B = 112.18$ s and of the sample C was $\theta_C = 283.30$ s. The computed values of parameter a and the local solidification time θ determine, via their ratio, the quotient of the diffusion coefficient D and the square of the structure parameter L, which means that the following relation applies:

$$a/\theta = D/L^2 \qquad [s^{-1}] \tag{13}$$

The calculated values of relation (13) for oxides of the samples B and C are arranged in Table 2 together with the parameters a.

Oxide	Sample B: α	$\alpha/\theta_B \cdot 10^4$ [1/s]	Sample C: α	$\alpha/\theta_C \cdot 10^4$ [1/s]
Na_2O	0.0732	6.53	0.0691	2.44
Al_2O_3	0.0674	6.01	0.0662	2.34
SiO_2	0.0741	6.61	0.0663	2.34
ZrO_2	0.00035	0.0312	0.00008	0.0028
K_2O	0.0721	6.43	0.0665	2.35
CaO	0.075	6.69	0.0703	2.48
TiO_2	0.0759	6.77	0.0757	2.67
Fe_2O_3	0.0732	6.53	0.0711	2.51
HfO_2	0.0165	1.47	0.00017	0.006

Table 2. Calculated values of the equation (13)

It comes as a surprise that the values of the parameter $\alpha/\theta = D/L^2$ of the oxides of elements Na, Al, Si, K, Ca, Ti, and Fe differed by as much as an order from the value of the same parameter of the oxide of zirconium and hafnium. This could be explained by the fact that zirconium contains hafnium as an additive and, therefore, they segregate together and the forming oxides of zirconium and hafnium show the highest melting temperatures. From the melt, both oxides segregated first, already in their solid states. Further redistribution of the oxides of both elements ran on the interface of the remaining melt and the successive segregation of other oxides only to a very limited extent. It was therefore possible to count on the fact that the real diffusion coefficients of zirconium and hafnium in the successively forming crystallites were very small (i.e. $D_{Zr} \to 0$ and $D_{Hf} \to 0$). On the other hand, the very close values of the parameters $\alpha/\theta = D/L^2$ of the remaining seven analyzed oxides:

$$D/L_B^2 = (6.51\pm0.25).10^{-4} \quad \text{and} \quad D/L_C^2 = (2.45\pm0.12).10^{-4} \quad [s^{-1}] \tag{14,15}$$

indicated that the redistribution of these oxides between the melt and the solid state ran in a way, similar to that within metal alloys, namely steels.

It would be possible to count – in the first approximation – the diffusion coefficients of the oxides in the slag having the temperatures of 1765°C (solidus) and 1775°C (liquidus), the average value of $D = (2.07\pm0.11) \times 10^{-6}$ cm^2/s (the data referred to the diffusion of aluminum in the slag of a composition of 39% CaO-20% Al_2O_3-41% SiO_2). For these cases, and using Equation (13), it was possible to get the magnitude of the structure parameters that governed the chemical heterogeneity of the values:

$$L_B = \sqrt{\left[\left(2.07\times10^{-6}\right)/\left(6.51\times10^{-4}\right)\right]} = 0.05639$$
$$L_C = \sqrt{\left[\left(2.07\times10^{-6}\right)/\left(2.45\times10^{-4}\right)\right]} = 0.09192 \qquad [cm] \tag{16,17}$$

It corresponded to 564 µm in the sample B (which was taken from the edge of the casting block) and 919 µm in the sample C (which was taken from underneath the riser of the same casting block). The comparison of the micro-structures of the analyses samples B and C (Figures 6a,b) has clearly shown that the sample B micro-structure (L_B) was significantly finer than the micro-structure of the sample C (L_C), which semi-quantitatively corresponded to the qualified estimate of the structure parameters L, conducted on the basis of calculations using the data obtained from both models.

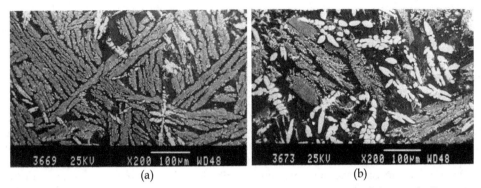

(a)   (b)

Fig. 6. a)The structure of the sample B (L_b = 564 μm), b) The structure of the ample C (L_c = 919 μm)

4. Continuously casting

4.1 Chemical microheterogeneity of continuously cast steel slab

Structure of metallic alloys is one of the factors, which significantly influence their physical and mechanical properties. Formation of structure is strongly affected by production technology, casting and solidification of these alloys. Solidification is a critical factor in the materials industry (Kavicka et al., 2007). Solute segregation either on the macro- or micro-scale is sometimes the cause of unacceptable products due to poor mechanical properties of the resulting non-equilibrium phases. In the areas of more important solute segregation there occurs weakening of bonds between atoms and mechanical properties of material degrade. Heterogeneity of distribution of components is a function of solubility in solid and liquid phases. During solidification a solute can concentrate in inter-dendritic areas above the value of its maximum solubility in solid phase. Solute diffusion in solid phase is a limiting factor for this process, since diffusion coefficient in solid phase is lower by three up to five orders than in the melt. When analysing solidification of steel so far no unified theoretical model was created, which would describe this complex heterogeneous process as a whole. During the last fifty years many approaches with more or less limiting assumptions were developed. Solidification models and simulations have been carried out for both macroscopic and microscopic scales. The most elaborate numerical models can predict micro-segregation with comparatively high precision. The main limiting factor of all existing mathematical micro-segregation models consists in lack of available thermodynamic and kinetic data, especially for systems of higher orders. There is also little experimental data to check the models. Many authors deal with issues related to modelling of a non-equilibrium crystallisation of alloys. However, majority of the presented works concentrates mainly on investigation of modelling of micro-segregation of binary alloys, or on segregation of elements for special cases of crystallisation – directional solidification, zonal melting, one-dimensional thermal field, etc. Moreover these models work with highly limiting assumption concerning phase diagrams (constant distribution coefficients) and development of dendritic morphology (mostly one-dimensional models of dendrites. Comprehensive studies of solidification for higher order real alloys are rarer. Nevertheless, there is a strong industrial need to investigate and simulate more complex alloys because nearly all current commercial alloys have many components often exceeding ten elements. Moreover, computer simulation have shown that even minute amounts of alloying elements can significantly influence microstructure and micro-segregation and cannot be neglected.

4.1.1 Methodology of chemical heterogeneity investigation

Original approach to determination of chemical heterogeneity in structure of poly-component system is based on experimental measurements made on samples taken from characteristic places of the casting, which were specified in advance. Next procedure is based on statistical processing of concentration data sets and application of the original mathematical model for determination of distribution curves of dendritic segregation of elements, characterising the most probable distribution of concentration of element in the frame of dendrite (Dobrovska et al., 2009), and the original mathematical model for determination of effective distribution coefficients of these elements in the analysed alloy.

4.1.2 Application of methodology of chemical heterogeneity investigation – investigation into chemical micro-heterogeneity of CC steel slab

A continuously cast steel slab (CC steel slab, Figure 7) with dimensions 1530x250 mm was chosen for presentation of results, with the following chemical composition in (wt. %): 0.14C; 0.75Mn; 0.23Si; 0.016P; 0.010S; 0.10Cr; 0.050Cu; 0.033Al$_{total}$.

After solidification and cooling of the cast slab a transversal band was cut out, which was then axially divided into halves. Nine samples were taken from one half for determination of chemical heterogeneity according to the diagram in Figure 8. The samples had a form of a cube with an edge of approx. 20 mm, with recorded orientation of its original position in the CC slab. Figure 9 shows an example of microstructure of the analysed slab. On each sample a concentration of seven elements (*aluminium, silicon, phosphor, sulphur, titanium, chromium and manganese*) were measured along the line segment long 1000 µm. The distance between the measured points was 10 µm.

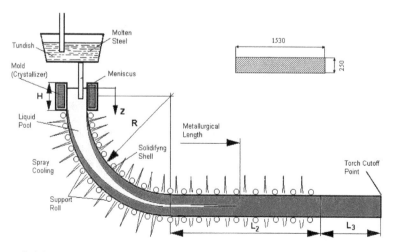

Fig. 7. The steel slab caster

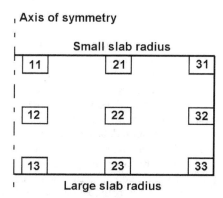

Fig. 8. Scheme of sampling from a slab and marking of samples

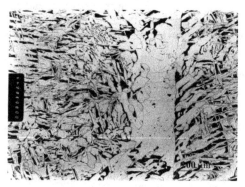

Fig. 9. Example of structure of the sample 21 with a microscopic   trace of 1000 µm long

Analytical complex unit JEOL JXA 8600/KEVEX Delta V Sesame was used for determination of concentration distribution of elements, and concentration was determined by method of energy dispersive X-ray spectral micro-analysis. As an example, Figures 10 a,b present *the basic concentration spectrum* of Mn, Si, P and S.

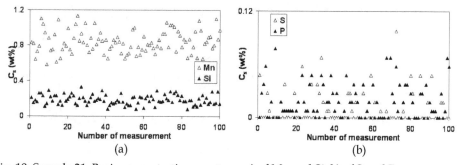

Fig. 10. Sample 21. Basic concentration spectrum a) of Mn and Si, b) of S and P

Chemical micro-heterogeneity, i.e. segregation of individual elements at distances, order of which is comparable to dendrite arms spacing, can be quantitatively evaluated from the basic statistical parameters of the measured concentrations of elements in individual samples. These parameters comprise: C_x average concentration of element (arithmetic average) in the selected section, s_x standard deviation of the measured concentration of element, C_{min} minimum concentration of element and C_{max} maximum concentration of element measured always on the selected section of the sample. It is possible to calculate from these data moreover *indexes of dendritic heterogeneity* I_H of elements in the measured section of individual samples as ratio of standard deviation s_x and average concentration C_x of the element. Then the element distribution profiles can be plotted according to the Gungor's method (Gungor,1989) from the concentration data sets measured by the method ED along the line segment 1000 μm long. Data plotted as the measured weight percent composition versus number of data (Figures 10 a,b) were put in an ascending or descending order and *x*-axis was converted to the fraction solid ($f_s \approx g_s$ in Equation 3) by dividing each measured data number by total measured data number. The element composition versus fraction solid, i.e. element distribution profile (*distribution curve of dendritic segregation*) was then plotted; Figures 11 a,b represent such dependences for manganese, silicon, phosphor and sulphur. The slope of such curve (ascending or descending) depended on the fact, whether the element in question enriched the dendrite core or the inter-dendritic area in the course of solidification.

From these statistical data it is also possible to determine with use of original mathematical model for each analysed element from the given set of samples the values of *effective distribution coefficients* k_{ef}. The procedure of the effective distribution coefficient calculation will be outlined here as follows:

The sequence of such arranged concentrations (Figures 11 a,b) was seen as a distribution of concentrations of the measured element in the direction from the axis ($f_S = 0$) to the boundary ($f_S = 1$) of one average dendrite.

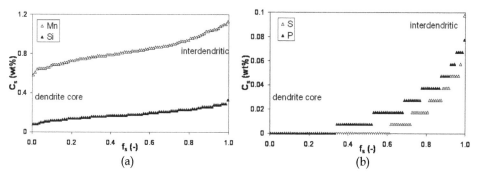

Fig. 11. Experimentally determined distribution curve of dendritic segregation (sample 21) a) for Mn and Si ,b) for P and S

The effective distribution coefficient k_{ef} was in this case defined by the relation

$$k_{ef}(f_s) = C_s(f_s) / C_L(f_s) \qquad (18)$$

where C_S is the solute concentration in the solidus and C_L is its concentration in liquidus and argument (f_S) expressed the dependence of both concentrations on the fraction solid. A perfect mixing of an element in the interdendritic melt was then assumed (this assumption.

It was therefore possible to substitute the equation (18) by the formula

$$k_{ef}(i) = C_i \, / \, C_R(i) \tag{19}$$

where C_i is the concentration in i-th point of the sequence (i.e. in the i-th point of the curve in Figures 11 a, b) and $C_R(i)$ is the average concentration of the element in the residual part of the curve (i.e. for $f_S \in \langle i, 1 \rangle$), expressed by the relation:

$$C_R(i) = \left[1 \, / \, (n - i + 1) \right] \sum_{j=1}^{n} C_j \tag{20}$$

where n was the number of the measured points. In this way it was possible to determine the values of effective distribution coefficients for all $i \in \langle 1, n \rangle$, i.e. for the entire curve characterising the segregation during solidification. The effective distribution coefficients of all the analysed elements were calculated by this original method. The average values of determined effective distribution coefficients are listed in Table 3. No segregation occurs when $k_{ef} = 1$; the higher is the deviation from the number 1, the higher is the segregation ability.

The effective distribution coefficients calculated in this way inherently include in themselves both the effect of segregation in the course of alloy solidification and the effect of homogenisation, occurring during the solidification as well as during the cooling of alloy.

Average values of measured and calculated quantities in the set of samples are in Table 4.

Sample		Element						
		Al	Si	P	S	Ti	Cr	Mn
11	I_H	1.24	0.28	1.22	1.45	0.30	0.22	0.14
	k_{ef}	0.32	0.78	0.33	0.26	0.76	0.83	0.88
12	I_H	1.54	0.30	1.12	1.74	0.29	0.27	0.15
	k_{ef}	0.24	0.77	0.36	0.20	0.78	0.79	0.88
13	I_H	1.44	0.30	1.25	1.48	0.30	0.29	0.15
	k_{ef}	0.27	0.78	0.32	0.26	0.77	0.78	0.88
21	I_H	1.33	0.29	1.58	1.49	0.31	0.24	0.13
	k_{ef}	0.29	0.78	0.24	0.25	0.76	0.81	0.89
22	I_H	1.14	0.28	1.31	1.41	0.30	0.26	0.14
	k_{ef}	0.35	0.78	0.30	0.27	0.77	0.80	0.88
23	I_H	1.56	0.29	1.34	1.86	0.26	0.28	0.13
	k_{ef}	0.24	0.78	0.29	0.18	0.80	0.78	0.89
31	I_H	1.11	0.28	1.22	2.34	0.31	0.23	0.16
	k_{ef}	0.37	0.78	0.33	0.18	0.76	0.82	0.87
32	I_H	1.44	0.27	1.16	1.49	0.34	0.25	0.14
	k_{ef}	0.27	0.79	0.34	0.25	0.74	0.80	0.88
33	I_H	1.32	0.29	1.24	1.64	0.35	0.26	0.13
	k_{ef}	0.30	0.78	0.32	0.22	0.74	0.80	0.89

Table 3. The average values of the heterogeneity index I_H and the effective distribution coefficient k_{ef} of elements in the individual samples

	c_x $\pm$ s_x	I_H $\pm$ s_I	k_{ef} $\pm$ s_k	$k^{(ref)}$ according to Dobrovska et al., 2009
Al	0.0136 0.0029	1.352 0.162	0.294 0.046	0.12 –0.92
Si	0.1910 0.0068	0.285 0.011	0.781 0.005	0.66 –0.91
P	0.0141 0.0023	1.270 0.133	0.314 0.035	0.06 –0.50
S	0.0136 0.0030	1.657 0.297	0.232 0.035	0.02 –0.10
Ti	0.0951 0.0032	0.306 0.027	0.765 0.019	0.05 –0.60
Cr	0.1758 0.0076	0.255 0.023	0.799 0.017	0.30 –0.97
Mn	0.8232 0.0169	0.143 0.009	0.873 0.033	0.72 –0.90

Table 4. Average values of the measured and calculated quantities in the set of all samples

Data represented in Table 3 and Table 4 make it possible to evaluate dendritic heterogeneity (micro-heterogeneity) of elements, as well as their effective distribution coefficients in individual samples, and also in the frame of the whole analysed half of the slab cross-section. It is obvious from these tables that dendritic heterogeneity of accompanying elements and impurities is comparatively high. This is demonstrated by the index of dendritic heterogeneity I_H. It follows from Table 3, that distinct differences exist between micro-heterogeneity of individual elements. Figures 12 a, b show distribution of indexes of micro-heterogeneity of sulphur (the most segregating element) and manganese (the least segregating element) on slab cross-section.

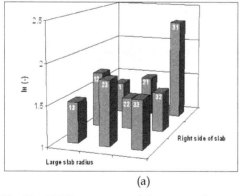

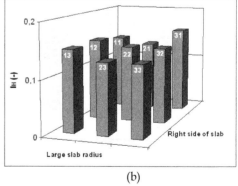

(a)   (b)

Fig. 12. a) Differences in sulphur micro-heterogeneity in samples taken from one-half of slab cross-section. b) Differences in manganese micro-heterogeneity in samples taken from one-half of slab cross-section.

Average value of this coefficient for all the analysed elements and the whole set of nine samples is given in Table 4. It follows from this table that dendritic heterogeneity of slab decreases in this order of elements: sulphur, aluminium, phosphor, titanium, silicon, chromium and manganese, which has the lowest index of heterogeneity. Dendritic heterogeneity of the analysed elements is expressed also by the values of their effective distribution coefficients, arranged for individual samples in Table 3 and for the set of samples in Table 4.

It is obvious from the tables that pair values of the index of dendritic heterogeneity and effective distribution coefficient for the same element do mutually correspond. The higher the value of the heterogeneity index, the lower the value of effective distribution coefficient and vice versa. The lowest value of the effective distribution coefficient is found in sulphur and the highest value is found in manganese. It follows from the Table 4, that effective distribution coefficient increases in this order of elements: sulphur, aluminium, phosphor, titanium, silicon, chromium and manganese. All the analysed elements segregate during solidification into an inter-dendritic melt, and their distribution coefficient is smaller than one. For comparison, the Table 4 contains also the values of distribution coefficients found in literature. It is obvious that our values of effective distribution coefficients, calculated according to the original model, are in good agreement with the data from literature, only with the exception of sulphur (and titanium). The reason for this difference is probably the means of calculation of the effective distribution coefficient – the value of this parameter is calculated from concentration data set measured on solidified and cooled casting. Consequently, the effective distribution coefficients calculated in this way inherently include in themselves both the effect of segregation in the course of alloy solidification and the effect of homogenisation, occurring during the solidification as well as during the cooling of alloy.

4.2 Effect of elelectromagnetic stirring on the dendritic structure of steel billets

Currently, casters use rotating stators of electromagnetic melt-stirring systems. These stators create a rotating magnetic induction field with an induction of **B**, which induces eddy-current **J** in a direction perpendicular to **B**, whose velocity is **v**. Induction **B** and current **J** create an electromagnetic force, which works on every unit of volume of steel and brings about a stirring motion in the melt. The vector product (**v** x **B**) demonstrates a connection between the electromagnetic field and the flow of the melt. The speeds of the liquid steel caused by the elelectromagnetic stirring is somewhere from 0.1-to-1.0 m/s. The stirring parameters are within a broad range of values, depending on the construction and technological application of the stirrer. The power output is mostly between 100 and 800 kW, the electric current between 300 and 1000 A, the voltage up to 400 V and with billet casting the frequency from 5 to 50 Hz.

The elelectromagnetic stirring applied on the steel caster is basically a magneto-hydraulic process together with crystallisation processes and solidification of billet steel. The complexity of the entire process is enhanced further by the fact that the temperatures are higher than the casting temperatures of concast steel. The temperature of the billet gradually decreases as it passes through the caster down to a temperature lying far below the solidus temperature. From the viewpoint of physics and chemistry, the course of the process is co-determined by a number of relevant material, physical and thermokinetic characteristics of the concast steel and also electrical and magnetic quantities. There is also a wide range of construction and function parameters pertaining to the caster and elelectromagnetic stirring as well as parameters relating to their mutual arrangement and synchronisation. Numerous works from recent years relate that exact mathematical modelling of elelectromagnetic stirring on a caster is still unsolvable (Stransky et al., 2009).

The basic elelectromagnetic stirring experiment was conducted on a continuously steel billet caster where two individual mixers were working (Figure 13). The first stirrer, entitled MEMS (Mould Electromagnetic Stirring), is mounted directly on the mould and the second stirrer, entitled SEMS (Strand Electromagnetic Stirring), is mounted at the beginning of the flow directly after the first cooling zones but in the secondary-cooling zone. Here the outer structure of the billet is already created by a compact layer of crystallites, however, in the centre of the billet there is still a significant amount of melt that is mixed by the SEMS.

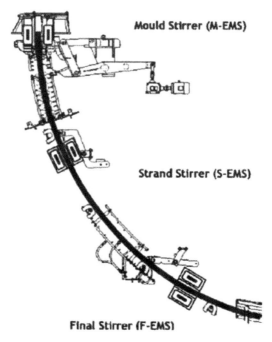

Fig. 13. The steel billet caster of 150x150 mm. The positions of the MEMS and SEMS stirrers

4.2.1 The temperature field of a billet

The temperature field of the billet of 150x150 mm computed via original numerical model (Stransky et al., 2009,2011) is in Figures 14-15.

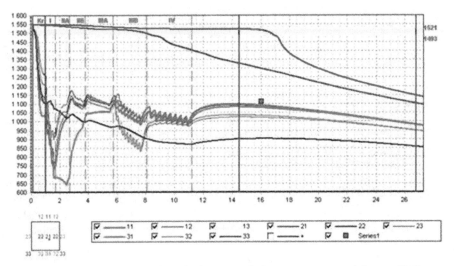

Fig. 14. The temperature history of marked points of the cross-section of the steel billet 150x150 mm

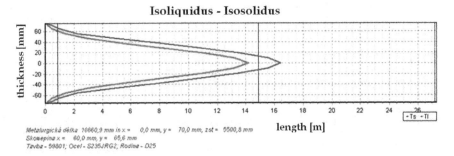

Fig. 15. Computed iso-solidus and iso-liquidus curves in the axial longitudinal section

4.2.2 The experiment

The first stirrer (MEMS) stirs the melt still in the mould while the billet is undergoing crystallization and solidification. The second stirrer (SEMS) works at a time when the melt is already enclosed by a shell of crystallites around the perimeter of the billet and inside the billet there is less melt than above in the active zone of the first stirrer. When both stirrers were switched off, the crystallisation and solidification continued in the normal way, i.e. the solidifying melt did not undergo a forced rotational movement. Samples were taken throughout the course of the experiment – from parts of the billet cast using the MEMS and SEMS and without and also using either one. The samples were taken in the form of cross-sections (i.e. perpendicular to the billet axis). The samples were fine-ground and etched with the aim of making visible the dendritic structure which is characteristic for individual variants of the solidification of the billet. The verification of the influence of MEMS and SEMS on the macrostructure of the billet was carried out on two melts of almost the same chemical composition (Table 5).

Melt	C	Mn	Si	P	S	Cu	Cr	Ni	Al	Ti
A	0.14	0.31	0.22	0.014	0.009	0.03	0.05	0.02	0.02	0.002
B	0.13	0.32	0.22	0.018	0.012	0.09	0.06	0.04	0.02	0.002

Table 5. Chemical composition of experimental melts [wt.%]

The timing of the concasting process of the billets – without the involvement of the stirrers and with the working of the elelectromagnetic stirring of individual variants of stirrers (MEMS and SEMS) – is given in Table 6. The speed of the concasting (i.e. the movement, the proceeding of the billet through the mould) of the billet was maintained constant during the experimentation at a value of 2.7 m/min. Table 6 shows that as many as nine concasting variants were verified. The lengths of individual experimental billets – from which samples had been taken – were always a multiple of the metallurgical length. The average superheating of the steel above the liquidus was 32.8 ± 3.1 °C in melt A and 28.0 ± 4.6 °C in melt B, which lies within the standard deviation of the temperature measurements.

Melt	Concasting mode – sampling	Superheating of steel above liquidus [°C]	MEMS stirring [Amperes]	SEMS stirring [Amperes]	Fig.
A	1A	37	210	0	
	2A	31	0	0	Fig. 16a
	3A	33	0	29	
	4A	30	210	57	Fig. 16b
B	1B	35	210	0	
	2B	30	0	0	
	3B	27	0	57	
	4B	24	210	57	
	5B	24	210	29	

Table 6. The billet concasting modes and sampling

Note: Detailed records of the experimental verification of the effects of MEMS and SEMS during concasting on the relevant device pertain to Table 6. The data are appended with a time history of the MEMS and SEMS connection and with information relating to the lengths of individual billets and the points from which the actual samples had been taken (i.e. the cross-sections from which the dendritic structures had been created). Evaluation of all nine variants of concasting (Table 6) indicates that the arrangement of dendrites in the cross-section follow the same tendency in the first phase of crystallization. The structure is created by columnar crystals – dendrites – perpendicular to the walls of the billet (Figure 16a).

In the billets that were not stirred the dendrites gradually touch one another on the diagonals of the cross-section. Here their growth either ceases, or the dendrites bend in the directions of the diagonals and their growth continues all the way to the centre of the billet. The columnar dendrites that grow from the middle part of the surface maintain their basic orientation – perpendicular to the surface – almost all the way to the centre of the billet. In the central part of the cross-section there is an obvious hollow on all nine macroscopic images. This is most probably a shrinkage.

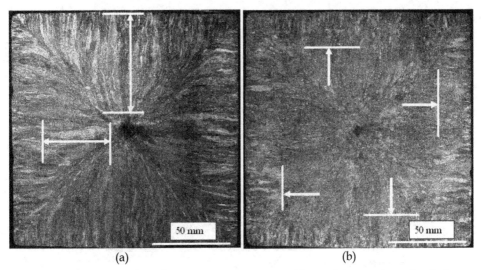

(a)   (b)

Fig. 16. a) Dendrite growth in the concasting structure without elelectromagnetic stirring – mode 2A, b) The growth of dendrites in the billet structure using the MEMS and SEMS – mode 4A (Table 6)

The above-described mechanism of dendrite growth during concasting without stirring is frequently the object of interest (Figure 16a). Inside the billets, when using the MEMS stirrer (or both MEMS and SEMS), the kinetics of solidification and dendrite growth is initially the same as without stirring. This also creates columnar dendrites which touch along the diagonals, however, soon their growth ceases still near the surface. Dendrites, which are called equiaxed dendrites continue to grow – their orientation is more random and only partly directed towards the centre of the billet (Figure 16b). It appears that this dendrite growth mechanism manifests itself the most when both stirrers are working simultaneously (Table 6: 4A, 4B and 5B). If MEMS and SEMS are working simultaneously, the stirring effect significantly destroys the formation of columnar crystals. If only MEMS is working and SEMS is switched off (1A and 1B), then the desctruction of columnar crystals is less evident. The working mode of SEMS alone (modes 3A and 3B) cannot be clearly differentiated from the changes in the dendritic structure in relation to the structure formed without stirring (2A and 2B).

Figure 16b (the macro-ground dendritic structure) shows the depth of the columnar band of dendrites in the direction away from the surface of the billet (Figure 16b – see arrows) and its value, which (with the simultaneous stirring of MEMS and SEMS) is 23.4±1.8 mm. The same qualified guess was made for ordinary billet casting (i.e. without stirring). Here, the depth of the dendrites can be guessed almost all the way to the central shrinkage at 70 mm (Figure 16a – see arrows). It is known that additives and impurities during solidification are often concentrated in points of contact of the growing dendrites, where the maximum of segregated additives and impurities and the greatest probability of technological defects occurs. In the given case, this undesirable effect can be expected along the diagonals which have a length of up to 100-to-103 mm towards the central shrinkage. This point of contact of the dendrites during the simultaneous working of SEMS and MEMS is only 29.8±1.9 mm, i.e. 3.4× less. The central area of the billet containing a hollow as a result of a shrinkage is then filled with dendrites growing into a vacuum (i.e. underpressure) (Figure 17).

Fig. 17. Dendrites in the centre of the billet

Under the assumption that the maximum of defects (i.e. vacancies, impurities, additives and micro-shrinkages) are formed along the diagonals it is possible to expect that in the areas of the corners – specifically on the edges – the nucleation of cracks will be higher than on the walls of the billet. If the first approximation of the fracture toughness of the relevant billet made from low-carbon steel is $K_{IC} \sim 75.0$ MPa.m$^{1/2}$, then in the ordinary concasting process it can be assumed that the length of the contact of columnar dendrites along the diagonal will be approximately $\Delta l_{normal} \approx 101.5$ mm (Figure 16a). On the other hand, if both electromagnetic stirrers (MEMS and SEMS) are engaged simultaneously, the contact length of the columnar dendrites along the diagonal decreases to $\Delta l_{el.magnl} \approx 29.8$ mm (Figure 16b). Along these lengths (i.e. the areas) it could be expected that during concasting the concentration of the primary defects will increase.

A comparison of limit stresses and strains in the area of the edges of the billets during concasting without elelectromagnetic stirring and if both MEMS and SEMS stirrers are engaged indicates that the billets (otherwise cast under the same conditions) cast without stirring are almost twice as susceptible to cracking along the edges as billets cast using both stirrers. A similar assumption can be made even in the case of assessing the effect of columnar dendrites in the central part of the surface of the billet where, without stirring, their length grows from the surface of the wall all the way to the central shrinkage (Figure 16a), while with the stirrers the dendrites are significantly shorter. The boundaries of the dendrites are however much less damaged by technological defects (vacancies, etc.) than the areas of their touching – of the peaks along the diagonals. Long-term statistical monitoring of the quality of 150×150 mm billets and the chemical composition has proven that the application of elelectromagnetic stirring has significantly reduced the occurrence of defects (in this case cracks).

5. Conclusion

Progressive creation of numeric model of unsteady thermal field A connected with the model of chemical heterogeneity B, leads to a completely novel conjugated numeric model, which necessarily requires respecting reality of poly-component crystallising metallic system, formed usually by eight to eleven constitutive elements. Constitutive elements forming conjugated model have during crystallisation completely different physical-chemical properties in dependence of temperature. Their redistribution in the volume of crystallising tangible macroscopic system is governed by the 2nd Fick's law. Mutual functional connection of both models A and B into one mutually cooperating conjugated model AB represents a completely new step resting on real crystallising poly-component system. This connection of two models AB necessarily requires large amount of consistent concentration data of constitutive elements forming real crystallising tangible macroscopic poly-component system. These data concern alloying elements (e.g. Ni, Cr), basic tramp elements (Mn, Si, Ti, V, Mo), data on admixture elements and impurities (S, P), as well as data on de-oxidising elements (Al, Ca, etc.).

The most complicated conjugated model will be the model for continuous casting. The authors have prepared for its creation 50,000 experimentally verified and mutually consistent data on elements. These data, make it possible to express concentrations C, effective distribution coefficient of elements between melt and crystallising solid phase k_{ef}, diffusion coefficients D_s in the melt of all segregating elements of dendritically crystallising system in the sense of the equation (3), express also the degree of heterogeneity and shares of solidified phase g_s. It contains also the equation (4), which is dimensionless Fourier's number α of the second kind for mass transfer, which contains implicitly, apart from the diffusion coefficient also share of local solidification time and squares of half distance of dendritic axes θ_{ls}/L^2. Equation (5) postulates by share of concentrations at the interface of the melt and solid phase (dendrite) the degree of heterogeneity – these mutually consistent and already verified data on elements form the basic starting point for progressive functional creation of the above mentioned conjugated model AB.

6. Acknowledgment

This research was conducted using a program devised within the framework of the GA CR projects GA CR projects No. 106/08/0606, 106/09/0370, 106/09/0940, 106/09/0969, P107/11/1566 and MSMT CR- MSM6198910013.

7. References

Boettinger, W.J. et al.(2000). Solidification microstructures: recent development, future directions, *Acta Mater.*, 48 pp. 43-70 ISSN 1359-6457

Brody, H.D. & Flemings, M.C. (1966). Solute redistribution in dendritic solidification, *Trans. AIME*, 236 (1966), pp. 615-624

Chalmers, B. (1964). *Principles of Solidification*. John Wiley & Sons, New York.

Chvorinov, N. (1954). *Krystalisace a nestejnorodost oceli [Crystallization and heterogeneity of steel]*. Nakladatelstvi CSAV, Praha

Davies, G.J. (1973). *Solidification and Casting*. Applied Science Pub., London.

Dobrovska, J., Stransky, K., Dobrovska, V. & Kavicka, F. (2009).Characterization of Continuously Cast Steel Slab Solidification by Means of Chemical Micro-heterogeneity Assessment, *Hutnicke listy* No.5, LXII , pp. 4-9, ISSN 0018-8069

Dobrovska, J., Kavicka F., Stransky, K., et al. (2010). Numerical Optimization of the Method of Cooling of a Massive Casting of Ductile Cast-Iron , *NUMIFORM 2010*, Vols 1 and

2 - Dedicated to professor O. C. Zienkiewicz (1921-2009) Book Series: AIP Conference Proceedings Vol. 1252 , pp. 578-585 , ISBN 978-0-7354-0799-2

Gungor, M. N. (1989) A statistically significant experimental technique for investigating micro-segregation in cast alloys, *Metall. Trans. A*, 20A, pp. 2529-2538.

Heger, J., Stetina, J., Kavicka, F., et al. (2002). Pilot calculation of the temperature field of the ceramic material EUCOR, *Advanced Computational Methods in Heat Transer VII* Book Series: Computational Studies, Vol. 4, pp. 223-232 ISBN 1-85312-906-2

Flemings, M. C. (1974). *Solidification Processing*. McGraw-Hill, New York.

Kavicka, F., Stetina, J., Sekanina, B., et al. (2007). The optimization of a concasting technology by two numerical models , *Journal of Materials Processing Technology* , Vol. 185, Issue 1-3, Sp. Iss. SI, pp. 152-159 , ISSN 0924-0136

Kavicka, F, Dobrovska, J., Sekanina, B., et al. (2010). A Numerical Model of the Temperature Fi.eld of the Cast and Solidified Ceramic Material , *NUMIFORM 2010*, Vols 1 and 2 - Dedicated to professor O. C. Zienkiewicz (1921-2009) Book Series: AIP Conference Proceedings , Vol. 1252 , pp. 571-577 ISBN 978-0-7354-0799-2

Kraft, T., Chang, Y. A. (1997) Predicting microstructure and microsegregation in multicomponent alloys, *JOM*, pp. 20-28.

Kurz, W., Fisher, D.J. (1986). *Fundamentals of Solidification*. Trans Tech Publications, Switzerland.

Rappaz, M. (1989). Modelling of microstructure formation in solidification processes, *Int. Mater. Rev.*, 34, pp. 93-123 ISSN 0950-6608

Smrha, L.(1983). *Tuhnutí a krystalizace ocelových ingotů* [*Solidification and crystallisation of steel ingots*] . SNTL, Praha

Stefanescu, D. M.(1995). Methodologies for modelling of solidification microstructure and their capabilities, *ISIJ Int.*, 35, pp. 637-650 ISSN 0915-1559

Stransky, K., Kavicka, F. , Sekanina, B. et al. (2009). Electromagnetic stirring of the melt of concast billets and its importance, *METAL 2009*, Hradec nad Moravici, ISBN 978-80-87294-04-8

Stransky, K., Dobrovska, J., Kavicka, F., et al. (2010). Two numerical models of the solidification structure of massive ductile cast-iron casting , *Materiali in Tehnologije / Materials and Technology* , Vol. 44, Issue 2, (MAR-APR 2010), pp. 93-98 , ISSN 1580-2949

Stransky, K., Kavicka, F., Sekanina, B. et al. (2011).The effect of electromagnetic stirring on the crystallization of concast billets, *Materiali in Tehnologije /Materials and Technology,* Vol. 45 , Issue 2, pp. 163–166, ISSN 1580-2949

3

Crystallization Kinetics of Chalcogenide Glasses

Abhay Kumar Singh

Department of Physics, Banaras Hindu University, Varanasi,
India

1. Introduction

1.1 Background of chalcogenides

Chalcogenide glasses are disordered non crystalline materials which have pronounced tendency their atoms to link together to form link chain. Chalcogenide glasses can be obtained by mixing the chalcogen elements, viz, S, Se and Te with elements of the periodic table such as Ga, In, Si, Ge, Sn, As, Sb and Bi, Ag, Cd, Zn etc. In these glasses, short-range inter-atomic forces are predominantly covalent: strong in magnitude and highly directional, whereas weak van der Waals' forces contribute significantly to the medium-range order. The atomic bonding structure is, in general more rigid than that of organic polymers and more flexible than that of oxide glasses. Accordingly, the glass-transition temperatures and elastic properties lay in between those of these materials. Some metallic element containing chalcogenide glasses behave as (super) ionic conductors. These glasses also behave as semiconductors or, more strictly, they are a kind of amorphous semi-conductors with band gap energies of 1±3eV (Fritzsche, 1971). Commonly, chalcogenide glasses have much lower mechanical strength and thermal stability as compared to existing oxide glasses, but they have higher thermal expansion, refractive index, larger range of infrared transparency and higher order of optical non-linearity.

It is difficult to define with accuracy when mankind first fabricated its own glass but sources demonstrate that it discovered 10,000 years back in time. It is also difficult to point in time, when the field of chalcogenide glasses started. The vast majority of time the vitreous glassy state was limited to oxygen compounds and their derivatives. Schulz-Sellack was the first to report data on oxygen-free glass in 1870 (Sellack, 1870). Investigation of chalcogenide glasses as optoelectronics materials in infra-red systems began with the rediscovery of arsenic trisulfide glass (Frerichs, 1950, 1953) when R. Frerichs was reported his work. Development of the glasses as a practical optoelectronic materials were continued by W. A. Fraser and J. Jerger in 1953 (Fraser et. al., 1953). During the 1950-1970 periods (Hilton, 2010) the glasses ware made in ton quantities by several companies and it frequently used in commercial devices. As an example, devices were made to detect the overheated bearings in the railroad cars. Hot objects could be detected by the radiation transmitted through the 3- to 5 µm atmospheric window, for this transparent arsenic trisulfide glass was used. While, to make chalcogenide glass compositions which capable in transmitting longer wavelengths arose the concept of passive thermal optical systems was adopted.

Jerger, Billian, and Sherwood (Hilton, 1966 & 2010) extended their investigation on arsenic glasses containing selenium and tellurium and later adding germanium as a third constituent. The goal was to use chalcogen elements heavier than sulfur to extend long-wavelength transmission to cover the 8- to 12 μm window with improve physical properties. A subsequent work also in Ioffee Institute, in Lenin-grad under the direction of Boris Kolomiets was also reported in 1959 (Hilton, 2010). Work along the same line was begun in the United Kingdom by Nielsen and Savage (Nielsen, 1962, Savage et. al., 1964, Savage et. al., 1966) as well as work also at Texas Instruments (TI) began as an outgrowth of the thermoelectric materials program. The glass forming region for the silicon-arsenic-tellurium system was planed by Hilton and Brau (Hilton et. al., 1963). This development led to an exploratory DARPA- ONR program from 1962 to 1965 (Hilton, 2010). The ultimate goal of the program was to find infrared transmitting chalcogenide glasses with physical properties comparable to those of oxide optical glasses and a softening point of 500°C. Futher, Hilton (Hilton, 1974) also worked on sulfur-based glasses in 1973 to 1974. The exploratory programs resulted the eight chalcogenide glass U.S. patents and a number of research paper published in an international journal detailing the results (Hilton et. al., 1966). After that scientific community made serious effort in development of chalcogenide glasses and organized symposia and meeting in All-Union Symposium on the Vitreous Chalcogenide Semiconductors held in May 1967 (Kokorina, 1966) in Leningrad (now called St. Petersburg). In that symposium Stanley Ovshinsky (founder of Electron Energy Conversion Devices in Troy, Michigan) was presented his first paper dealing switching devices based on the electronic properties of chalcogenide glasses. Similar work was reported by A. D. Pearson (Pearson, 1962) in United States, thus started a great world wide effort to investigate chalcogenide glasses and their electronic properties. The purpose was to pursue a new family of inexpensive electronic devices based on amorphous semiconductors. The effort in this field far exceeded the effort directed toward optoelectronics applications. Some of the results of the efforts in the United States were reported in a symposium (Doremus, 1969) and in another symposium (Cohen et. al.,1971). In this order Robert Patterson mapped out the glass forming region for the germanium-antimony-selenium system and granted a U.S. patent (Patterson, 1966-67) covering the best composition selection.

In 1967 Harold Hafner was made many important contributions including a glass casting process and a glass tempering process in Semiconductor Production Division under the direction of Charlie Jones. There work concentrated the efforts on a glass from the germanium-arsenic-selenium system and outcomes (Jones et. al., 1968) agreed with the conclusions of the Russian, U.K. Alternatively, Servo group efforts that the germanium-arsenic-selenium system produced the best glasses for infrared system applications. Don Weirauch (Hilton, 2010) was conducted a crystallization study on the germanium-arsenic selenium family of glasses and identified a composition in which crystallites would not form. In 1972 a commercial group was successfully cast (12 in 24 in 0.5 in) a window which flat polished, parallel and antireflection-coated (Hafner, 1972). In the late 1960s and early 1970s, passive 8- to 12μ m systems began to be produced in small numbers mostly for the defense uses.

In 1968, Ovshinsky and his co-workers was discovered (Stocker, 1969) the some chalcogenide glasses exhibited memory and switching effects. After this discovery it became clear that the electric pulses could be switch the phases in chalcogenide glasses back and forth between amorphous and crystalline state.Around the same period in 1970's, Sir N. F. Mott (a former Noble Prices winner in Physics-1977) and E.A. Davis were developed the theory on the electronic processes in non-crystalline chalcogenide glasses (Mott et. al., 1979), and Kawamura (Kawamura et. al., 1983) was discovered xerography. Applications of solar cells were developed by Ciureanu and Middehoek (Ciureanu et. al., 1992) and Robert and his coworkers (Robert et.al., 1998). Infrared optics applications were studied by Quiroga and Leng and their coworkers (Quiroga et al., 1996, Leng et. al., 2000). The switching device applications were introduced by Bicerono and Ovshinsky (Bicerono et. al.,1985) and Ovshinsky (Ovshinsky, 1994). P. Boolchand and his coworkers (Boolchand et.al., 2001) was discovered intermediate phase in chalcogenide glasses. In this order several investigators have been also reported that the useful optoelectronics applications in infrared transmission and detection, threshold and memory switching (Selvaraju et al., 2003), optical fibers (Bowden et al., 2009, Shportko et al., 2008, Milliron et al.,2007) functional elements in integrated-optic circuits (Pelusi et al.2009) non-linear optics (Dudley et al., 2009), holographic & memory storage media (Vassilev et al., 2009, Wuttig et al.,2007), chemical and bio-sensors (Anne et al.,2009, Schubert et at., 2001), infrared photovoltaics (Sargent, 2009), microsphere laser (Elliott, 2010), active plasmonics (Samson,2010), microlenses in inkjet printing (Sanchez, 2011) and other photonics (Eggleton, 2011) applications. In this respect, the analysis of the composition dependence of their thermal properties was an important aspect for the study (Singh et al., 2009, 2010, 2011).

Subsequently, several review books were published on chalcogenide glasses e. g. "The Chemistry of Glasses" by A Paul in 1982, "The Physics of Amorphous Solids" by R.Zallen in 1983 and "Physics of Amorphous Materials" by S.R.Elliott in 1983. However, first book entirely dedicated to chalcogenide glassy materials entitled "Chalcogenide Semiconducting Glasses" was published in 1983 by Z.U.Borisova. In this order, G.Z. Vinogradova was published her monograph "Glass formation and Phase Equilibrium in Chalcogenide Systems" in 1984. M.A. Andriesh dedicates a book to some specific applications of chalcogenide glasses entitled "Glassy Semiconductors in Photo-electric Systems for Optical Recording of Information". M.A. Popescu gave large and detailed account on physical and technological aspect of chalcogenide systems in his book "Non-Crystalline Chalcogenides". The compendium of monographs on the subject of photo-induced processes in chalcogenide glasses entitled "Photo-induced Metastability in Amorphous Semiconductors" was compiled by A. Colobov-2003. Robert Fairman and Boris Ushkov-2004 described physical properties in "Semiconducting Chalcogenide Glass I: Glass formation, structure, and simulated transformations in Chalcogenide Glass". Finally, A. Zakery and S.R. Elliott demonstrated the "Optical Nonlinearities in Chalcogenide Glasses and their Applications" in 2007.

1.1.1 Binary chalcogenides

Structure of chalcogenide glasses have been extensively studied in binary compositions considering both the bulk and thin film forms. Chalcogens can form alloys together, Se-S, Se-Te (Gill, 1973) and S-Te amorphous (Sarrach et. al, 1976, Hawes, 1963) compounds were identified; however the scientific community seems to have, for the moment at least, left these glasses aside. Many binary compounds can be synthesized by associating one of the chalcogen with another element of the periodic table like, indium, antimony, copper, germanium, phosphorus, silicon and tin. A few other compounds based on heavy or light elements and alkali atoms have also been investigated. Abrikosov and his co-workers in 1969 (Lopez, 2004) were first reported the molecular structures of most extensively studied As-S, As-Se binary chalcogenide alloys in their monograph, the phase diagrams for the As- S and As-Se systems. As-S alloys can be formed with an As content up to 46%, while in As-Se this maximum content can be raised to almost 60%. Glasses with low As content can easily crystallize (e.g. for a content of 6% As the glass crystallizes at room temperature in one day) in the range 5-16 weight %, in a couple of days at 60°C while it takes 30 days for As S at 280°C (Lopez, 2004). As-Se alloys can crystallize along the all composition range, however this was to be done under pressure and at elevated temperatures. The typical $As_2 S_3$ structure has usually pictured as an assembly of six AsS pyramids (the As atom being the top of the pyramid while three S atoms form the base). Goriunova and Kolomiets in 1958 (Lopez, 2004) were pointed out that the importance of covalent bonding in chalcogenide glasses as the most important property to make stability of these glasses. As opposed to metallic bonding, covalent bonding ensures easier preparation of the glasses. Thus, the crosslinking initiated by the As atoms should reduce the freedom for disorder in which bonds are covalent. Further Vaipolin and Porai-Koshits reported X-ray studies in beginning of the 1960's (Lopez, 2004), for the vitreous $As_2 S_3$ and As_2Se_3 and a number of binary glass compositions based on these two compounds. These glasses were shown to contain corrugated layers, which deformed with increasing size of the chalcogens and arsenic atoms became octohedrally coordinated. The character of the bonds was also found become more ionic when at equimolecular compositions. At the beginning of the 1980's, Tanaka was also characterized the chalcogenide glasses as a phase change materials and demonstrated that they structurally rigid and not having long-range ordering.

Alternatively, Se-S, Se-Te and S-Te (Hamada et. al., 1968, Bohmer & Angell, 1993) were extensively studied binary chalcogens alloys in which Se and S taken as host material. Amorphous selenium and sulfer molecular structure become the mixture of chain and rings which bridging the gap between molecular glasses and polymers. They covalently bonded with two coordination number. The most stable trigonal structural phase a-Se consists of parallel helical chains and two monoclinic bonds forms the composed of rings of eight atoms. These polymorphs distinguished by the correlation between neighboring dihedral angles. The amorphous selenium has relatively low molecular weight polymer with low concentration of rings (Bichara & pellegatti, 1993, Caprion & Schober, 2000 & 2002, Echeveria et. al., 2003, Malek et. al., 2009, etc).

Particularly Se-In binary chalcogenide compositions were getting much attention due to their versatile technical applications. VI- III family compounds Se- In form layered structures with strong covalent bonds. Basically VI-III group Se- In compounds have hexagonal symmetry structure. It consists of two layers which separated by tetrahedrally or pentagonally coordinated Se and In (JablÇonska et. al., 2001 & Pena et al., 2004). Amorphous Se-In compounds contains a-In_2Se_3, a-$InSe$, and a-In_4Se_3 binary phases. The number of In-Se nearest neighbor heteropolar bonds considerably larger than the homopolar bonds. The In-In and Se-Se homopolar bonds contribute mainly to the left- and right-hand side of the first peak in the radial distribution function, but they do not influence original position. The number of nearest-neighbor Se-Se bonds in a-$InSe$ and a-In_4Se_3 structures is generally negligible (Kohary et. al., 2005).

1.1.2 Ternary chalcogenides

Ternary chalcogenide glasses also broadly studied from more than three decades. Ternary chalcogenides can be prepared by introducing a suitable additive element in well known or new binary matrix. Most extensively studied ternary As-S-Se system was shown a very wide glass-forming region (Flaschen et. al., 1959). The solid solutions can be formed along the line $As_2 S_3$ –$As_2 Se_3$ which proved via IR spectra and X-ray analysis by Velinov and his coworkers (Velinov et. al., 1997). The Covalent Random Network (CRN) and the Chemically Ordered Network (CON) models both satisfy the 8-N rule under the distribution of bond types in a covalent network with multi elements. As- rich glasses can be formed As–As, As–Se, and As–S bonds; thus Se-rich glasses have As–Se, As–S, and Se–Se bonds and S-rich glasses As–Se, As–S, and S–S bonds. The relative weight of each of the above units is expected to be proportionate to the overall composition of the glass itself (Yang et. al., 1989).

In recent years Zn containing ternary chalcogenide glasses attracted much attention due to higher melting point, metallic nature and advanced scientific interest (Boo et. al., 2007). Crystalline state zinc has hexagonal close-packed crystal structure with average coordination number four. While, in amorphous structure it is expected to metallic Zn dissolve in Se chains and makes homopolar and heteropolar bonds. Addition of third element concentration in binary alloy affects the chemical equilibrium of exiting bonds, therefore newly form ternary glass stoichiometry would heavily cross-linked, and makes homopolar and heteropolar bonds in respect of alloying elements. Specifically, Se-Zn-In ternary chalcogenide glasses can form Se-In heteronuclear bonds with strong fixed metallic Zn-In, Zn-Se bonds. Incorporation of indium concentration as-cost of selenium amount, the Se-Zn-In became heavily cross-linked results the steric hindrance increases at the threshold compositional concentration and beyond the threshold concentration a drastic change in physical properties has been observed.

1.1.3 Multicomponent chalcogenides

Addition of more than three elements in chalcogenide alloys refers as multicomponent alloys. In recent years there is an intensive interest made on study of new multi-component chalcogenide glasses to make sophisticated device technology as well as from the point of view of basic physics. Although Se rich binary and ternary chalcogenide glasses exhibit high resistivity, greater hardness, lower aging effect, enhanced electrical and optical properties with good working performance. But ternary glasses have certain drawbacks which implying the limitation in applications. It is worth then to add more than two components into selenium matrix can produce considerable changes in the properties complex glasses. Predominantly, metal and semimetal containing multi-component amorphous semiconductors promising materials to investigations such as; Ge-Bi-Se-Te, Al-(Ge-Se-Y), Ge–As–Se–Te, Cd (Zn)- Ge(As), $GeSe_2$– Sb_2Se_3–PbSe, $Cu_2ZnSnSe_4$ etc. (Thingamajig et. al., 2000, Petkov, 2002, Vassilev, 2006, Wibowo et. al. 2007). More specifically Se-Zn-Te-In multi-component chalcogenide glasses make Se-In heteronuclear bonds with other possible bonds Zn-In and Te-In. Due to the addition of Indium in quaternary glassy matrix, the structures become heavily cross-linked and steric hindrance increases. Therefore, at the expanse of Se chains and replacement of weak Se-Se bonds by Se-In bonds results the increase and decrease in their associative physical properties of Se-Zn-Te-In glasses.

Therefore, the thermal, electrical and optical properties of chalcogenide glasses widely depend on alloying concentration and intrinsic structural changes make them a chemical threshold a particular concentration of alloy. In view of these basic property several past research work in chalcogenide glasses were reported on binary, ternary and very few on multicomponent systems (Tonchev et. al., 1999, Wagner et. al., 1998, Mehta et. al., 2008, Patial et. al., 2011, Malek et. al., 2003, Soltan et. al., 2003, Song et. al., 1997, Usuki et. al., 2001, Fayek et. al., 2001, Wang et. al., 2007, Vassilev et. al., 2007, Eggleton et. al., 2011, Prashanth, et. al., 2008, Vassilev et. al.,2007, Othman et. al., 2006, Zhang et. al.,2004, Hegab et. al., 2007, Narayanan et. al., 2001, etc). Scientific and technological drawbacks, like low thermal stability, low crystallization temperature and aging effects (Guo et. al., 2007, Boycheva et. al., 2002, Ivanova et. al., 2003, Vassilev et. al., 2005, Xu et. al., 2008, Troles et. al., 2008) of non-metallic binary and ternary alloys motivates to investigators to make metallic multicomponent chalcogen alloys to achieve high thermal stability and harder chalcogenide glasses (Pungor, 1997, Demarco et. al., 1999, Kobelke et. al.,1999, Zhang et. al., 2005, Singh, 2011).

Extensive research on metal containing multicomponent chalcogenide alloys was begin nearly end of nineties when Kikineshy and Sterr 1989 & 1990 (Kokenyesi, et. al.,2007 & Ivan & Kikineshi, 2002) were demonstrated the multilayer of chalcogen alloys simply nanostructures materials which can be rather easily produced with controlled geometrical parameters. In this order Ionov and his coworkers (Ionov et. al.,1991) were demonstrated the electrical and electrophotographic properties of selenium based metal containing multicomponent chalcogenide glass and outlined these materials would be useful for electrophotographic and laser printer photoreceptors. Saleh and his coworkers (Saleh et. al., 1993) were studied the nuclear magnetic resonance relaxation of Cu containing chalcogenide glasses. Carthy & Kanatzidis (Carthy & Kanatzidis, 1996) were introduced the bismuth and antimony containing new class of multicomponent chalcogenide glasses. In the same year Natale and his coworkers (Natale et. al.,1996) were demonstrated the heavy metal multicomponent glasses useful for array sensors. Further, Nesheva and his coworkers (Nesheva et. al., 1997) were studied the amorphous pure and allying selenium based multilayers and demonstrated the photoreceptor properties at room temperature unaltered throughout in a year. Efimov (Efimov,1999) was described the mechanism of formation of the vibrational spectra of glasses such as quasi-molecular model, central force model and its recent refinements (model of phonon localization regions) and deduce the trends in the IR and Raman band assignments in inorganic systems. Goetzberger & Hebling (Goetzberger & Hebling, 2000) were commented on the present, past and future of photovoltaic materials. In the same year Naumis (Naumis, 2000) demonstrated the jump of the heat capacity in chalcogenide glasses during glass transition and show change in glass fragility and excess thermal expansivity is a function of average coordination number. While Mortensen and his coworkers (Mortensen et. al., 2000) were used heavy metals based chalcogenides sensors in detection of flow injection. Mourizina and his coworkers (Mourizina et. al., 2001) were demonstrated the ion selective light addressable poentiometric senor based on metal containing chalcogenide glass film. In the same year Rau and his coworkers (Rau et. al., 2001) were studied the effect of the mixed cation in chalcogenide glasses and reported that the non-linear structural changes in Raman and infrared spectra. Further, Messaddeq and his coworkers (Messaddeq et. al., 2001) were demonstrated the light induced volume expansion in chalcogenide glasses under the irradiation UV light. Moreover, Hsu and Narayanan and their coworkers (Hsu et. al., 2001, Narayanan et. al., 2001) were studied the near field microscopic properties of electronic & photonic materials and devices and large switching fields in metal containing chalcogenide glasses owing to chemical disordering. Salmon & Xin (Salmon & Xin, 2002) were studied the effect of high modifier content corresponding to coordination number and demonstrate the structural motifs change in such materials. Subsequently, Agarwal & Sanghera (Agarwal & Sanghera, 2002) were discussed the development and application of chalcogenide glass optical fibers in near scanning field microscopy/spectroscopy and Jackson & Srinivas (Jackson & Srinivas, 2002) demonstrated the modeling of metallic chalcogenide glasses using density function theory calculations. Tanaka (Tanaka, 2003) reviewed the nanoscale structures of chalcogenide glasses and inspect surface modifications at nanometer resolution and Micoulaut & Phillips (Micoulaut & Phillips ,2003) were shown the three elastic phases of covalent networks ((I) floppy, (II) isostatically rigid, and (III) stressed-rigid) depend on the degree freedom of material. They were also suggested that the ring factor is responsible for high crystallization temperature in metallic/ semi-metallic chalcogenide glasses. Lezal and his coworkers (Lezal et. al., 2004) were reviewed the chalcogenide glasses for optical and photonics applications. Vassilev & Boycheva (Vassilev & Boycheva, 2005) were critically reviewed the achievements in application of chalcogenide glasses as membrane materials. They were also demonstrated that the advantages and disadvantages in analytical performance and compared with the corresponding polycrystalline analogous. Emin (Emin, 2006) was explained the polaron conduction machines in amorphous semiconductors and Kokenyesi (Kokenyesi, 2006) reviewed the amorphous chalcogenide nano-multilayers: research and developments. While Phillips (Phillips, 2006) demonstrated the, ideally glassy materials have hydrogen-bonded networks. Bosch and his coworkers (Bosch et. al. 2007) were critically reviewed the last decade developments in optical fibers in bio sensing and Vassilev and his coworkers (Vassilev et. al.,2007) introduced the new Se- based multicomponent chalcogenide glasses and studied their composition dependence physical properties. Furthermore, Wachter and Taeed their coworkers (Wachter et. al., 2007, Taeed et. al., 2007) were demonstrated the composition dependence reversible and tunable glass-crystal-glass phase transition properties in new class of multicomponent chalcogenide glasses and show chalcogenide glasses useful for all-optical signal processing devices due to their large ultrafast third-order nonlinearities, low two-photon absorption and the absence of free carrier absorption in a photosensitive medium. Dahshan and Lousteau and their coworkers (Dahshan et. al.,2008, Lousteau et. al.,2008) were demonstrated the thermal stability and activation energy of some Cu doped chalcogenide glasses and the fabrication of heavy metal fluoride glass to explore the optical planar waveguides by hot-spin casting. Further, Klokishner (Klokishner et. al., 2008) were studied the concentration effects on the photoluminescence band centers in multicomponent metallic chalcogenide glasses and Ielmini and his coworkers (Ielmini et. al., 2008) demonstrated the threshold switching mechanism by high-field energy gain in the hopping transport of chalcogenide glasses. Mehta and his coworkers (Mehta et. al., 2009) were studied the effect of metallic and non metallic additive elements on Se-Te based chalcogenide glasses. In the same year Turek and Anne their coworkers (Turek et. al.,2009, Anne et. al.,2009) were demonstrated the artificial intelligence/fuzzy logic method for analysis of combined signals from heavy metal chemical sensors and commented on, due to the remarkable properties of chalcogenide glasses can be used as a biosensor which can collect the information on whole metabolism alterations rapidly. Further, Khan and his coworkers (Khan et. al., 2009) were demonstrated the composition dependence electrical transport and optical properties of metallic element doped Se based chalcogenide glasses. In order to this Snopatin and his coworkers (Snopatin et. al., 2009) were demonstrated the some high purity multicomponent chalcogenide glasses for fiber optics. Kumar and his coworkers (Kumar et. al., 2010) were demonstrated the calorimetric studies of Se-based metal containing multicomponent chalcogenide glasses. Peng & Liu (Peng & Liu ,2010) were reviewed the advances and achievements in SPM-based data storage in viewpoint of recording techniques including electrical bistability, photoelectrochemical conversion, field-induced charge storage, atomic manipulation or deposition, local oxidation, magneto-optical or magnetic recording, thermally induced physical deformation or phase change, and so forth as well as achievements in design and synthesis of organic charge-transfer (CT) complexes towards thermochemical-hole-burning memory, the correlation between hole-burning performances and physicochemical properties of CT complexes.

Story of the investigations will be remain continue in field of metallic chalcogenides (not limited to above outlined the major events) to deduce the new future prospective multicomponent chalcogenide glassy alloys. Amorphous chalcogenide alloys which full fill the essential requirement of modern optoelectronics. So, it can be outlined potential field of optoelectronics and advanced material is rapidly growing owing to their possible uses. Therefore, it is important to have an understanding regarding on crystallization process of chalcogenide glasses (predominately in metal containing alloys).

1.2 Crystallization

Crystallization is a natural process of formation of solid crystals from a solution/- melt. Crystallization of a substance can also achieve from the chemical solid-liquid separation technique, in which mass transfer from the liquid solution to a pure solid crystalline phase. The crystallization process of a substance mainly consists of two major events nucleation and crystal growth. In nucleation process the molten molecules dispersed in solid solution and begin to formation of clusters at the nanometer scale. Crystal growth is the subsequent growth of the nuclei which develop critical size of the formed clusters (because size of clusters plays an important role in the application of the material). Hence, the nucleation and growth are the continuous process which occurs simultaneously when supercooling exists in a system. Thus, system supercooling state acts as a driving force for the crystallization process. The supercooling driving force depending upon the conditions, either nucleation or growth may be predominant over to other and outcomes can be formed crystals with different sizes and shapes. Once the supercooling is established in a solid-liquid system and reached at equilibrium then crystallization process is completed (Mersmann, 2001).

In general supercooled materials/ or alloys have ability to crystallize with different crystal structures, this process is known as polymorphism. Each polymorph is in fact a different thermodynamic solid state and crystal polymorphs of the same alloy/-compound which exhibited different physical properties, such as dissolution rate, shape, melting point, etc. Thus the crystallization process of a substance is governed by both thermodynamic and kinetic factors which highly variable and difficult to control. Factors those affect the crystallization process of a substance/alloy are the impurity level, mixing regime, vessel design, and cooling profile and shape of crystals. Usually in those materials crystallization process occurs at lower temperatures, in supercooling situation they obey the law of thermodynamics. Its literal meaning a crystal can be more easily destroyed than it is formed.

Subsequently it is easier to a perfect crystal in a molten solid than to grow again a good crystal from the resulting solution. Hence the nucleation and growth process of a crystal are well control under the thermodynamic kinetic.

1.3 Nucleation and growth

1.3.1 Nucleation

Nucleation of the substance reflects the initiation of a phase change in a small region cause the formation of a solid crystal from a liquid solution. It is a consequence of rapid local fluctuations on a molecular scale in a homogeneous phase which define as a metastable equilibrium state. The whole nucleation process of a substance is the sum of heterogeneous (nucleation that occurs in the absence of a second phase) and heterogeneous (nucleation that occurs in the presence of a second, foreign phase) category of nucleation. Homogeneous nucleation due to clustering of molecules (embryos) in a supersaturated environment, in which a process began, combines two or more than molecules. In the reversible clustering process a few molecules grew at the same time and others dissolving. Once embryos attained a certain critical size then it decrease its total free energy by growing and becomes stable (Reid et. al., 1970). But, in practice it is difficult to find complete homogeneous nucleation owing to presence of insoluble amounts of matter even in pure material. Therefore, heterogeneous nucleation is always associated with homogeneous nucleation due to presence of second phase in bulk molten material. The heterogeneous nucleation occurs in a random fashion at various sites in a matter.

1.3.2 Crystal growth

As earlier mentioned crystal growth is the successive process of nucleation in which the critical nuclei of microscopic size form a crystal. Crystal growth in crystallization process takes place by fusion and re-solidification of the material. In this process within a solid material constituent of molecules (embryos) are arranged in an orderly repeating pattern extending in all three spatial dimensions. Crystal growth is a major stage of a crystallization process which consists the addition of new molecules (embryos) strings into the characteristic arrangement of a crystalline lattice. The growth typically follows an initial stage of either homogeneous or heterogeneous nucleation. The crystal growth process yields a crystalline solid whose molecules are typically close packed with fixed positions in space relative to each other. In general crystalline solids are typically formed by cooling and solidification from the molten (or liquid) state. As per the Ehrenfest classification it is first-order phase transitions with a discontinuous change in volume (and thus a discontinuity in the slope or first derivative with respect to temperature, dV/dT) at the melting point. Hence, the crystal and melt are distinct phases with an interfacial discontinuity having a surface of tension with a positive surface energy. Thus, a metastable parent phase represents it always stable with respect to the nucleation of small embryos from a daughter phase with a positive surface of tension. Hence, crystal growth process is first-order transitions consist advancement of an interfacial region whose structure and properties vary discontinuously from the parent phase. In the crystal growth process stiochiometry of glass compositions do not undergo in compositional changes during crystallization, mean, no need to long-range diffusion (Swanson, 1977) for crystal growth in chalcogenide glasses; thus, interfacial rearrangements are likely to control the crystal growth process. This (melt quenched) type of crystal growth is generally described from these three basic standard models: (i) the screw dislocation model; (ii) the normal or continuous growth model; and (iii) the two-dimensional surface nucleation growth.

Theoretically, nucleation and crystal growth process of molten solids first reported by Volmer and Weber (Volmer and Weber, 1925) and later on it explained in a large amount of literature. Tammann (1925) discussed the theory of nucleation and crystal growth and outlined the various parameters involved in terms of a probabilistic model involving a functional relation to pressure, temperature, and time. Many later studies deal with the nucleation and growth of crystals on an atomistic level. Nucleation of crystals from a melt is the mobility of atoms and molecules in the melt as measured by the diffusion coefficient. In glass-forming systems, liquid diffusion coefficients drop markedly with decreasing temperature (Towers and Chipman, 1957). When temperature drops below the liquidus, nucleation will increase from zero to a maximum at some undercooling. Diffusion rates then very low and nucleation decreases with further decrease in temperature. This explains the pattern of nucleation in liquids where the liquid diffusivity decreases with temperature and finally it grow a crystal. Rate of growth is thus a function of mobility of crystal-forming species within the melt. Mobility, can be measured by the diffusion coefficient, as drops with decreasing temperature and growth rate, like nucleation.

Thermodynamics of crystal growth process in molten solids can be expressed as (Warghese, 2010); the thermodynamical equilibrium between solid and liquid phases occur when the free energy of the two phases are equal

$$G_L = G_S \tag{1}$$

here G_L, and G_S are representing solid and liquid phases free energies

Free energy, internal energy and entropy of a system can be related from the Gibbs equation

$$G = H - TS \tag{2}$$

here G is the Gibbs free energy, H is the enthalpy, S is the entropy and T is the temperature.

Formation of a crystal can be considered as a controlled change of phase to the solid state. Therefore, the driving force for crystallization comes from the lowering of the free energy of the system during this phase transformation. Change in free energy in such transition can be related as;

$$\Delta G = \Delta H - T\Delta S \tag{3}$$

Where $\Delta H = H_L - H_S$

$$\Delta S = S_L - S_S$$

$$\Delta G = G_L - G_S$$

For the equilibrium $\Delta G = 0$

$$\Delta H = T_e . \Delta S$$

Where, T_e is the equilibrium temperature

$$\Delta G = \Delta H. \, \Delta T / T_e \tag{4}$$

Where $\Delta T = T_e - T$

ΔG is positive when $T_e > T$ and it depends on the latent heat of transition. The change in free energy can also represent as the product of change in entropy and super cooling temperature ΔT.

$$\Delta G = \Delta S. \, \Delta T \tag{5}$$

Although equation (5) representing melt growth, in which one may depend on concentration rather than supercooling for solution growth and vapour growth. Equation (5) in more convent form can be expressed as;

$$\Delta G \sim RT \ln (C/C_0) \tag{6}$$

$$\Delta G \sim RT \ln (P/P_0) \tag{7}$$

In general form

$$\Delta G \sim RT \ln S \tag{8}$$

where R is the Rydberg constant, C, C_0 are the concentration of solid solution and concentration at critical transition, P, P_0 are the pressers in vapour phase and S is the supercooling ratio.

Thus the equations (4) and (6) explain how the free energy changes depend on the supercooling parameters which are decisive in the process of crystallization. The rate of growth of a crystal can be expressed as a monotonically increasing function of Gibbs free energy, when other parameters remain the same.

Since in nucleation process in the supercooled solution forms the small clusters of molecules (Joseph, 2010), therefore, free energy change between the solid and liquid can be expressed as ΔG_v and Gibbs equation can be written as;

$$\Delta G = 4\pi r^2 \sigma - 4/3 \, \pi r^3 \, \Delta G_v \tag{9}$$

where σ- is the interfacial energy and r is the spherical radius of the molecule

The surface energy increase in term of r^2 and volume energy decreases in term of r^3. The critical size of nucleus can be obtained by the equation (9)

$$r^* = 2 \, \sigma / \, \Delta G_v \tag{10}$$

Hence the critical size (**r***) of the nucleus decreases with increasing cooling rate.

Further, in formation of critical nucleus size free energy (**ΔG^***) change can be calculated as;

$$\Delta G^* = 16\pi\sigma^3 / \, 3\Delta G_v^2 \tag{11}$$

Equation (11) in terms of Gibbs thermodynamical relation can be written as;

$$\Delta G^* = 16\pi\sigma^3 \Omega / \, 3(kT \ln S)^2 \tag{12}$$

where Ω is the molecular volume

The rate of nucleation (J) can be expressed as;

$$J = J_0 \exp [-\Delta G^*/ kT] \qquad (13)$$

or in terms of themodynamical parameters

$$J = J_0 \exp [-16\pi\sigma^3\Omega^2/ 3k^3T^3 (\ln S)^2] \qquad (14)$$

where J_0 is the pre-exponential factor

For critical supercooling condition J = 1, so that ln.J = 0, then expression can be expressed as;

$$S_{cri} = \exp [16\pi\sigma^3\Omega^2/ 3k^3T^3 \ln J_0]^{1/2} \qquad (15)$$

1.4 Crystallization kinetics

Study of the crystallization of the amorphous materials with respect to time and temperature is called crystallization kinetics. Crystallization kinetic study of the materials can be performed in either isothermal or non-isothermal mode of Differential Scanning Calorimetry (DSC). In the isothermal method, the sample is brought near to the crystallization temperature very quickly and the physical quantities, which change drastically are measured as function of time. In the non-isothermal method the sample is heated at a fixed rate and physical parameters recorded as a function of temperature. Investigators (Sbirrazzuoli, 1999) preferred to perform DSC measurements in non-isothermal mode. Owing to fact, it is not possible to ensure the homogeneity (or constant) of DSC furnace temperature in isothermal mode during the injection of material sample.

Crystallization kinetics parameters of the materials generally interpreted at glass transition temperature (T_g), crystallization temperature (T_c) and peak crystallization temperature (T_p) with help well defined statistical approximations (such as Hurby, Ozawa, Augis and Bentt, Moynihan and Kissinger) (Hruby,1972, Ozawa, 1970, Augis & Bennett, 1978, Moynihan et al., 1974, Kissinger,1957). All the existing approximations described on the basis of JMA (Johnson,1939, Avrami, 1939& 1940) model statics, although in recent years investigators also reported the (Sanchez-Jimenez, 2009) JMA model not a universal model to explain the crystallization kinetics of the materials, because it has few limitations. Despite of this majority view of investigators toward to kinetic methods based on JMA model are more reliable to explain the crystallization of chalcogenide glasses, polymers, metallic and oxide glasses.

1.5 Theoretical basis of crystallization kinetics

In isothermal phase transformation the extent of crystallization (α) of a certain material can be represent from the Avrami's equation (Moynihan et al., 1974, Kissinger,1957, Johnson,1939)

$$\alpha(t)=1-\exp[(-Kt)^n] \qquad (16)$$

where K is the crystallization rate constant and n is the order parameter which depends upon the mechanism of crystal growth.

In general the value of crystallization rate constant K increases exponentially with temperature. The temperature dependence behaviour of K indicates that the crystallization of amorphous or amorphous glassy materials is a thermally activated process. Mathematically it can be expressed as

$$K = K_0 \exp(-E_c / RT) \tag{17}$$

Here E_c is the activation energy of crystallization, K_0 is the pre-exponential factor, R is the universal gas constant and T is the temperature. For isothermal condition parameters E_c and K_0 in Eq. (17) can be assumed practically independent of the temperature (at least in the temperature interval accessible in the calorimetric measurements).

In non-isothermal crystallization, it is assumed that the constant heating rate during the experiment. The relation between the sample temperature T and the heating rate β can be written as

$$T = T_i + \beta t \tag{18}$$

Here T_i is the initial temperature. The crystallization rate is obtained by taking the derivative of expression (1) with respect to time, keeping in mind that the reaction rate constant is a time function which represents Arrhenius temperature dependence.

$$(d\alpha/dt) = n(Kt)^{n-1}[K + (dK/dt)t](1-\alpha) \tag{19}$$

The derivative of K with respect to time can be obtained from Eqs. (17) and (18), which follows as:

$$(dK/dt) = (dK/dT)(dT/dt) = (\beta E_c/RT^2)K \tag{20}$$

From Eq. (19) we obtained

$$(d\alpha/dt) = nK^n t^{n-1}[1+at].(1-\alpha) \tag{21}$$

where $a = (\beta E_c/RT^2)$.

Using Eq. (21) Augis and Bennett (Augis & Bennett, 1978) have developed a crystallization kinetic method. They have taken proper account of the temperature dependence crystallization reaction rate. Their approximation results have also been verified the linear relation between $\ln(T_c - T_i)/\beta$ versus $1/T_c$ (here T_c is the onset crystallization or critical transition temperature). This can be deduced by substituting the u for Kt in Eq. (21), accordingly the rate of reaction can be expressed as

$$(d\alpha/dt) = n(du/dt)u^{(n-1)}(1-\alpha) \tag{22}$$

Where

$$(du/dt) = u[(1/t) + a] \tag{23}$$

Second derivatives of Eqs. (22) and (23) are given as:

$$(d^2\alpha/dt^2) = [(d^2u/dt^2)u - (du/dt)^2 \times (nu^2 - n + 1)]nu^{(n-2)}(1-\alpha) = 0 \tag{24}$$

$$(d^2u/dt^2) = (du/dt)[(1/t) + a] + u[(-1/t^2)] + (da/dt) \tag{25}$$

In Eq. (25) substituting for $(da/dt) = -(2\beta/T)a$, then it can be written as

$$(d^2u/dt^2) = u\,[a^2 + (2a\,T_i/\,tT)] \tag{26}$$

The last term in the above equation was omitted in the original derivation of Augis and Bennett [33] (T_i T) and resulted in the simple form:

$$(d^2u/dt^2) = a^2\,u \tag{27}$$

Substitution of (du/dt) and (d^2u/dt^2) in Eqs. (23), (27), and (24) gives the following expression:

$$(nu^n-n+1) = [at/(1+at)]^2. \tag{28}$$

For E/RT >> 1, the right-hand bracket approaches its maximum limit and consequently u (at the peak) = 1, or

$$u = (Kt)_c = K_0\exp(-E_c\,/\,RT_c)\,[\,(T_c-T_i)\,/\,\beta] \approx 1 \tag{29}$$

In logarithm form, for $T_i \ll T_c$

$$\ln(\beta/T_c) \approx (-E_c\,/\,RT_c) + \ln K_0 \tag{30}$$

Value of E_c and K(T) can be obatined from the equation (30) by using the plots of $\ln\,\beta/T_c$ against $1/T_c$. Further, by using the Eq (17) Hu et.al. (Sbirrazzuoli, 1999) have introduced the crystallization rate constant stability criterion corresponding to T_c.

$$K\,(T_c) = K_0\exp(-E_c/RT_c\,) \tag{31}$$

1.6 Differential Scanning Calorimetry (DSC) thermograms

Endothermic and exothermic peaks in amorphous glassy materials arise due to thermal relaxation from a state of higher enthalpy toward to metastable equilibrium states of lower enthalpy. The process of the thermal relaxation depends on temperature and may quite fast near the glass-transition temperature. The glass transition peak in DSC measurement represents the abrupt change in specific heat and decrease in viscosity (Matusita, 1984), while the crystallization peak demonstrate to the production of excess free-volume, and endothermic peak at T_m reflects the amount of energy which liberate owing to complete destroy the solid phase structure cause braking of all type existing bonds in solid alloy. Hence, the materials crystallizations temperatures as well as mode of crystallizations extensively depend on the compositions of alloys.

In general DSC thermograms of amorphous glassy (i.e. chalcogenide glasses) materials have exhibited a considerable shifts in endothermic glass-transition and exothermic crystallization temperatures with increasing heating rates. But in recent investigations (Singh & Singh, 2009) investigators have also been reported vary small or negligible endothermic glass transitions shifts in metal, semi-metal and non metal containing multicomponent chalcogenide glasses. In order to this, we have performed the DSC measurements on recent developed $Se_{93-x}Zn_2Te_5In_x$ ($0\leq x \leq 10$) chalcogenide glasses.

These materials could be prepared by the well known most convenient melt quenched method. The high purity elements Selenium, Zinc, Tellurium and Indium were used. The suitable amounts of elements were weighed by electronic balance and put into clean quartz ampoules (length of ampoules 8 cm and diameter 14 mm). All the ampoules were evacuated and sealed under at a vacuum of 10^{-5} Torr to avoid the reaction of glasses with oxygen at high temperature. A bunch of sealed ampoules was heated in electric furnace up to 1173K at a rate of 5-6 K/min and held at that temperature for 10-11 h. During the melting process ampoules were frequently rocked to ensure the homogeneity of molten materials. After achieving desired melting time, the ampoules with molten materials were frequently quenched into ice cooled water. Finally ingots of glassy materials were obtained by breaking the ampoules. The preparation and characterizations technique of the test materials also outlined in our past [Singh & Singh, 2010] research work.

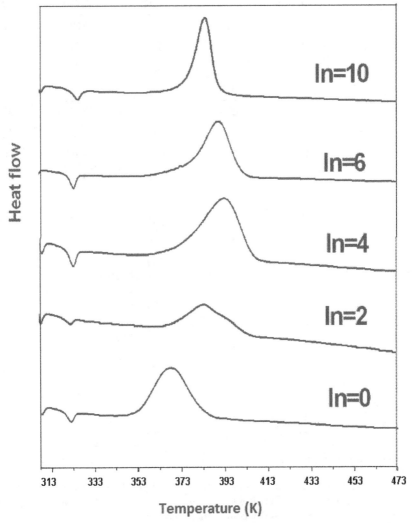

Fig. 1. DSC patterns of Se_{93-x}-Zn_2-Te_5-In_x ($0 \le X \le 10$) chalcogenide glasses at heating rate 5 K/min

DSC patterns of $Se_{93-x}Zn_2Te_5In_x$ ($0 \leq x \leq 10$) glasses at heating rate of 5 K (min)$^{-1}$ is given in Figure.1. DSC traces clearly show the endothermic and exothermic phase reversal peaks at the glass transitions and crystallizations temperatures. Obtained values of the glass transitions temperatures (T_g), onset crystallizations temperatures (T_c), peak crystallizations temperatures (T_p) and melting temperatures (T_m) at heating rates of 5, 10, 15 and 20 K (min)$^{-1}$ is listed in Table 1.

	Se_{93-x}-Zn_2-Te_5-In_x (X=0, 2, 4, 6 and 10)				
	Heating rate	T_g (K)	T_c (K)	T_p (K)	T_m(K)
Se_{93}-Zn_2-Te_5	5	318	354	368	501
	10	320	359	374	502
	15	322	363	378	503
	20	323	367	381	503
Se_{91}-Zn_2-Te_5-In_2	5	319	370	385	502
	10	322	376	392	503
	15	323	382	398	504
	20	324	385	402	504
Se_{89}-Zn_2-Te_5-In_4	5	319	372	393	503
	10	322	381	402	504
	15	324	385	407	504
	20	325	389	411	504
Se_{87}-Zn_2-Te_5-In_6	5	320	375	396	504
	10	323	383	405	504
	15	325	389	412	506
	20	327	392	415	506
Se_{83}-Zn_2-Te_5-In_{10}	5	319	370	386	504
	10	322	378	394	504
	15	323	383	400	504
	20	324	386	405	505

Table. 1 Obatined values of T_g, T_c, T_p and T_m at heating rates 5, 10, 15 and 20 K/min

Outcome demonstrates, a very small glass-transitions temperatures shifts and a considerable shifts in onset and peak crystallizations temperatures, the corresponding T_c-T_g result is given in Figure.2 and their values listed in Table.2. This result revealed the values of T_c and T_p crystallizations temperatures increases upto 6 at. wt. % indium and beyond this decreased for 10 atomic percentage composition glass.

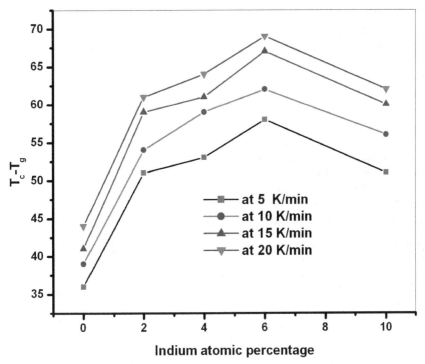

Fig. 2. Variations of T_c-T_g with Indium atomic percentage at 5, 10, 15 and 20 K/min heating rates

	Heating rate	$Se_{93}Zn_2Te_5$	$Se_{91}Zn_2Te_5In_2$	$Se_{89}Zn_2Te_5In_4$	$Se_{87}Zn_2Te_5In_6$	$Se_{83}Zn_2Te_5In_{10}$
	5	36	51	53	55	51
	10	39	54	59	60	56
T_c-T_g	15	41	59	61	64	60
	20	44	61	64	65	62
	5	14	15	21	21	16
	10	15	16	21	22	16
T_p-T_c	15	15	16	22	23	17
	20	14	17	22	23	19
	5	147	132	131	126	134
	10	143	127	123	119	126
T_m-T_c	15	140	122	119	114	121
	20	136	119	115	110	119

Table 2. Evaluated values of T_c-T_g, T_p-T_c and T_m-T_c at different heating rates

1.7 Glass forming ability (GFA)

GFA of material describes the relative ability to a set of compound adopt the amorphous structure (Mehta et al., 2006, Jain et al., 2009). In practice criteria to establish the GFA of vitreous materials are based on DSC measurements. Usually, unstable glass has show a crystallization peak near to the glass transition temperature while stable glass peak close to melting temperature. GFA can be evaluated by meen of the difference between the crystallization temperature (peak temperature T_p and/or onset temperature T_c) and the glass transition temperature (T_g). This difference varies with alloys concentrations and higher and lower for a certain composition. To evaluate the GFA of glassy alloys several quantitative methods have been introduced from the investigators. Most of the methods (Saad & Poulin,1987, Dietzel,1968) based on characteristics temperatures of glassy alloys. Dietzel (Dietzel,1968) has introduced the first GFA criterion $D_T = T_c - T_g$. Further Hruby (Hruby,1972) developed the H_R GFA criterion [$H_R = T_c-T_g / T_m-T_c$].This method has additional advantage to describe the thermal stability of amorphous materials.

To applying the GFA criterion method, obtained critical characteristics temperatures difference T_c-T_g, T_p-T_c and T_m-T_c (Here T_g is glass transition temperature, T_p is peak crystallization temperature and T_m is the melting temperature) values of under examine materials is listed in Table 2. Using these values H_R parameter of GFA can be described as:

$$H_R = \left(\frac{T_c - T_g}{T_m - T_c} \right) \tag{32}$$

GFA variation with indium atomic weight percentage at heating rates 5, 10, 15 and 20 is given in Figure. 3 and their corresponding average values listed in Table 3. The higher GFA value is obtained for threshold indium concentration glass. High GFA value of threshold composition also reflects their high order thermal stability as compare to other glasses of this series.

1.8 Activation energy

Activation energy reflects the involvement of molecular motions and rearrangements of the atoms around the critical transitions temperatures (Suri et al., 2006). In DSC measurement atoms undergo infrequent transitions between the local (or metastable state) potential minima which separated from different energy barriers in the configuration space, where each local minima represent a different structure. The most stable configuration has local minima structure in glassy region. This literal meaning a glass atoms possessing minimum activation energy have a higher probability to jump in metastable state of lower internal energy configuration. This local minima configuration occurs at particular composition of alloy which refers as a most stable glass (Imran et al., 2001). The activations energies of chalcogenide glasses at the critical temperatures can be interpreted in these words: the glass transition activation energy (E_g), onset crystallization activation energy (E_c) and peak crystallization activation energy (E_p) are the amount of energies which absorbed by a group of atoms for a jump from one metastable state to another state (Imran et al., 2001, Agarwal et al., 1991

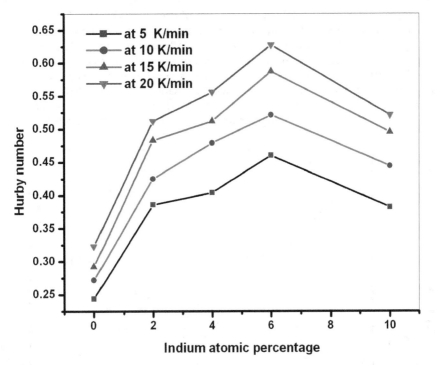

Fig. 3. Plots of GFA parameter with indium atomic percentage at 5, 10, 15 and 20 K/min heating rates

1.8.1 Glass transition activation energy

The glass transition activation reflects the endothermic energy of the material which produced due to unsaturated or hydrogen like bond braking at the pre-crystallization critical temperature. Glass transitions activations energies of the under test glasses can be defined by using Ozawa method (Ozawa, 1970).

$$\ln \beta = -\left(\frac{E_g}{RT_g} \right) + C \tag{33}$$

where β is the heating rate, E_g is the glass transition activation energy, C is the constant in usual meaning. Obtained Ozawa plots, $\ln \beta$ vs $1000/T_g$, for these materials is given in Figure.4 and their corresponding E_g values is listed in Table 3.

Outcomes revel the E_g values have a maxima and a minima for 0 and 6 percentage indium compositions glasses. Thus the E_g values have very small increasing and decreasing trend in T_g values (see Table 3&Table 1) with increasing DSC heating rates in these metal, semi-metal and non-metal containing multicomponent chalcogenide glasses. The high activations energies at pre-crystallizations reflect the materials rigidity. While, normally reported T_g values for non-metallic chalcogenides compositions has show a considerable shifts with

DSC heating rates. Hence the obtained T_g values results in increasing DSC heating rates for metal, semi-metal and non-metal containing multicomponent chalcogenide glasses are not in good agreement with previous reported non metallic compositions. Deviations in the results arise due to existence relatively hard metallic, semi-metallic characters unsaturated bonds with hydrogen like week bonds in the alloys stoichiometrics. Further, it is quite possible to large amount of metallic, semi-metallic characters unsaturated bonds sustain over to T_g critical transition temperature of the materials owing to requirement greater amount of energy to bark the heteropolar unsaturated bonds. As consequence the metal,

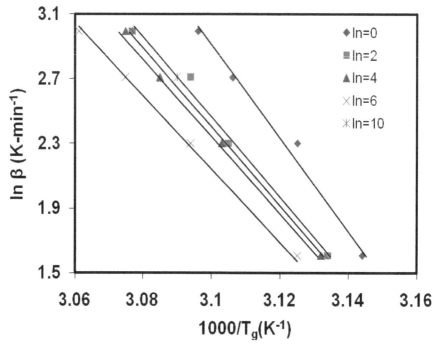

Fig. 4. Ozwa polts of $Se_{93-x}-Zn_2-Te_5-In_x$ ($0 \leq X \leq 10$) glasses to obtain E_g

Se₉₃₋ₓ-Zn₂-Te₅-Inₓ (X=0, 2, 4, 6 and 10)				
Alloy compositions	Glass transition activation energy E_g (KJ/mol)	Onset crystallization activation energy E_c (KJ/mol)	Peak crystallization activation energy E_p (KJ/mol)	Average (GFA)
$Se_{93}Zn_2Te_5$	229.94	115.64	118.23	0.282
$Se_{91}Zn_2Te_5In_2$	205.02	106.11	96.40	0.451
$Se_{89}Zn_2Te_5In_4$	195.47	99.02	91.09	0.487
$Se_{87}Zn_2Te_5In_6$	174.23	93.66	82.77	0.548
$Se_{83}Zn_2Te_5In_{10}$	199.88	101.84	92.70	0.460

Table 3. E_g, E_c, E_p and GFA values of $Se_{93-x}-Zn_2-Te_5-In_x$ ($0 \leq x \leq 10$) chalcogenide glasses

semi-metal and non-metal containing multicomponent chalcogenide glasses have exhibited either very small or negligible glass transitions temperatures shifts with increasing DSC heating rates.

1.8.2 Onset crystallization activation energy

Onset crystallization activation energy (E_c) is the amount of thermal energy which requires to begin the phase transformation from glassy to crystallization state. Quantitive knowledge of onset crystallization activation energy at T_c defines the heat/energy storage capability of the material which useful for different physical applications. The exothermic onset crystallization activation energy at T_c arises due to barking of existing covalent bonds in glassy configuration. In case of complex metallic multicomponent chalcogenide glasses high energy homopolar and heteropolar covalent bonds formed as compare to metallic binary and ternary compositions. Due to this the critical onset crystallization temperatures of the complex metallic glasses (see Table 1&Table 3) increases and their corresponding activation energies tend to be decrease upto threshold composition then visa-verse direction. While in case of non-metallic binary, ternary and multicomponet chalcogenide alloys reports demonstrated they have lower values of onset crystallizations temperatures owing to existence of week homopolar and heteropolar bonds in glassy configuration.

Onset crystallizations activations energies of under examine complex metallic multicomponent chalcogenide glasses described by employing the Ozawa method (Ozawa, 1970).

$$\ln \beta = -\left(\frac{E_c}{RT_c}\right) + C \tag{34}$$

Here symbols (β is heating rate, E_c is the onset crystallization activation energy, T_c is the onset crystallization temperature and R & C are the constant) are in usual meaning. Obtained Ozawa plots ln β vs $1000/T_c$ is given in Figure. 5 and their corresponding E_c values is listed in Table 3. Outcomes show a phase reversal in E_c values which have a maxima and minima corresponding to 0 and 6 atomic weight percentage of indium glasses.

1.8.3 Peak crystallization activation energy

Peak crystallization activation energy (E_p) of a glass expresses the amount of heat energy which requires for utmost crystallization. By mean at peak crystallization point almost all the existing heteropolar covalent bonds have to be broken and material achieve to maximum crystallization i.e. a glassy phase material completely transform to crystalline phase and relax toward to original state. The E_p values of examined materials can be described by using the Kissinger method (Kissinger,1957).

$$\ln\left(\frac{\beta}{T_P^2}\right) = -\frac{E_P}{RT_P} + C \tag{35}$$

Here symbols (β is heating rate, E_p is the peak crystallization activation energy, T_p is the paek crystallization temperature and R & C are the constant) are in usual meaning. Obtained E_p values from Kissinger plots (see Figure. 6) are listed in Table.3.

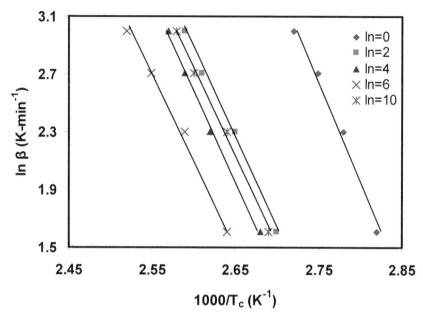

Fig. 5. Ozwa polts of $Se_{93-x}-Zn_2-Te_5-In_x$ ($0 \leq X \leq 10$) glasses to obtain E_c

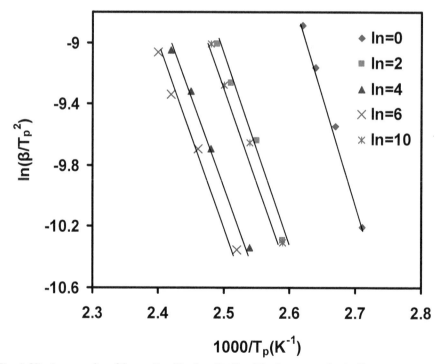

Fig. 6. Kissinger polts of $Se_{93-x}-Zn_2-Te_5-In_x$ ($0 \leq X \leq 10$) glasses to obtain E_p

Values of E_p also show a phase reversal with alloying compositions and have a maxima and minima respectively for 0 and 6 atomic weight of indium. Commonly, in metal, semimetal and non-metallic elements containing multicomponent chalcogenides show a sharp and continuous crystallization process (exception is also reported in few composition of chalcogenide glasses) with lower E_p values. The sharp crystallization prevail between T_c(where crystallization began) and T_p (where crystallization completed) owing to continuous braking of rigid heteropolar bonds cause generation of greater amount of heat energy in the specimen.

1.9 Melting temperature (T_m)

Melting temperature of amorphous glassy materials defines as; temperature at which solid state materials destroy all the existing homopolar and heteropolar bonds and alloying elements separated. Melting temperatures of amorphous glassy materials extensively depend on the constituent of the alloys. Technologically kinetics at T_m have less impotence, therefore investigators interest to provide only introductory information regarding to phase transformation at T_m in amorphous glassy materials.

2. Discussions

Crystallizations kinetics variations in under test metal, semi-metal and non-metal containing multicomponent chalcogenide glasses can be interpreted in term of bond formation in solids. It is expected to Zn and Te dissolved in Se chains and makes Zn-Zn, Te-Te, Se-Se, Se-Zn, Se-Te, Se-Zn-Te homopolar and heteropolar bonds. Essentially ternary Se-Zn-Te glass can forms cross-link heteropolar metastable state structure. The heteropolar bonds will be produced the defects in density of localized state owing to existence of dangling bonds in alloy configuration (Maharjan et al., 2000, Saffarini, 2002, Abdel Latif, 1998). Further incorporation of foreign element Te concentration in ternary configuration transforms the whole stoichiometry into quaternary or multicomponent system. The metal, semi-metal and non-metal multicomponent glassy configuration possibly makes them dominating Se-In heteropolar bonds with other metallic character bonds Zn-In, Te-In. The Se-In heteropolar bonds play an important role in crystallization kinetics variations due to fixed amounts of Zn and Te. Addition of additional indium concentration has produced the heavily cross-linked structure in which steric hindrance increases. Therefore the expanse of Se chains and replacement of weak Se-Se bonds by Se-In bonds results the increase and decrease in associative activations energies. A chemical threshold has established at critical composition (6 at wt % of In). At this concentration glassy structure become more chemically ordered and contains large number of Se-In bonds (Singh, 2011). As consequence a significant change is appeared in crystallization parameters of threshold composition alloy.

Furthermore, incorporation of indium concentration beyond the threshold composition reduced the Se-In bonds and increases the In-In bond strength in glassy configuration. The increase and decrease bonds strengths of Se-In and In-In influenced the defects / dangling bonds concentrations in the glassy stoichiometry. Owing to alternation in dangling bonds densities the GFA, activations energies E_g, E_c and E_p of the corresponding glass show a significant change in kinetic parameters.

3. Summary

In summary, in this work an effort is made to present the fundamentals (in short form) of nucleation and growth processes and crystallization in amorphous glassy (chalcogenide glasses) materials. Further, project a clear view on natural crystallization and process of non–isothermal crystallization kinetics of chalcogenide glasses. Subsequently a concrete explanation on origin of endothermic and exothermic peaks in DSC measurements is also discussed. Glass forming ability and crystallizations kinetics of recent developed $Se_{93-x}Zn_2Te_5In_x$ ($0 \leq x \leq 10$) metal, semi-metal and non-metal containing multicomponent chalcogenide glasses have also been taken under discussion. Outcomes revealed such combinations glasses have high order GFA and thermal stability and low E_c, E_p activations energies as compare to previous reported non-metallic compositions. In subsequent it has found these materials critical kinetic parameters extensively depend on their alloys constituents and have a maxima and a minima in respective manor for threshold composition glass. It is also concluded that the crystallizations kinetics variations in these glasses can be occur owing to fluctuations in solid state bonds densities in localized states. In case of metallic multicomponent chalcogenide glasses heteropolar unsaturated and covalent bonds may play an important role in crystallization kinetics variations.

Indeed, chalcogenide glasses are the potential materials which used in various optoelectronics applications, but still have plenty of room for their applications in different areas which have either less studied or undiscovered. A few thrust areas of these materials are outlined here in which they have (or may have) potential applications. Surface plasmon resonance (SPR) is a very versatile and accurate technique to determining small changes in optoelectronics parameters like, refractive index at the interface of a metal layer and the adjacent dielectric medium. The SPR detection mechanism has secured a very important place among several sensing techniques due to its better performance and reliable procedure. Chalcogenide materials can also perform as infrared sensing for an efficient, non-destructive and highly selective technique in detection of organic and biological species. This technique has combined benefits of ATR spectroscopy with the flexibility of using a fiber as the transmission line of the optical signal, which allows for remote analysis during field measurements or in clinical environments. Sensing mechanism based on absorption of the evanescent electric field, which propagates outside the surface of the fiber and interacts with any absorbing species at the fiber interface. However, their efficiency controlled multilayer optical filters by periodically switching and evaporation angle, leading to periodic dielectric structure makes them a potential candidate for chemical sensing application. Further, their larger refractive index amongst the glasses makes them to made chalcogenide based ultra-low loss waveguides devices. Chalcogenide materials have most promising applications in area of phase change memory (PCM) and in photonics. Their unique physical characteristic the reversible amorphous to crystalline phase change which can be induced by controlled thermal cycling (through laser absorption or current flow) in certain chalcogenide alloys. Phase-change materials have always technological importance to make read–write storage device (commercially rewritable CD/DVD), because they can be switched (in nanoseconds) rapidly back and forth between amorphous and crystalline phases by applying appropriate laser heat pulses. Although optical phase-change storage is a widespread and successful technology, further advances in areal densities will be very challenging. Moreover, chalcogenide (in glassy or nano embedded) based photovoltaic cells applications are also identified as a prominent alternative of conventional energy which can provides terawatts capacity at cheaper cost. But improvement in low efficiency chalcogenide based photovoltaics is challenging in future. Thus, in view of author glassy and nano embedded (glassy) chalcogenides would be matter of future research to undersatnd the molecular /- or nano-phase photonics of the materials, particularly for thrust areas in PCM memory and photovoltaics applications. In author opinion chalcogenide glasses (bulk or nano phase) has a bright future and it is still open for further inventions.

4. Acknowledgment

Author thankful to Dr.Kedar Singh and senior faculty members of Department of Physics, Banaras Hindu University, Varanasi-221005, India, for their kind support to carry out this work.

5. References

Abdel Latif, R. M. (1998). DC Electrical measurements on evaporated thin films of vanadium pentoxide. *Physica B*,Vol. 254,pp. 273-276, ISSN 0921-4526

Agarwal, P., Goel, S., Rai, J. S. P., Kumar, A. (1991). Calorimetric studies in glassy $Se_{80-x}Te_{20}In_x$, *Phys Stat Sol (a)*, Vol. 127, pp.363-369,ISSN 1862-6319

Aggarwal, I. D. & Sanghera, J. S. (2002). Development and applications of chalcogenide glass optical Fibers at NRL. *Journal of Optoelectronics and Advanced Materials*, Vol. 4, pp.665 – 678, ISSN 2066-0049

Anne, M.L., Keirsse, J., Nazabal, V., Hyodo, K., Inoue, S., Pledel, C.B., Lhermite, H., Charrier, J., Yanakata, K., Loreal, O., Person, J. L., Colas, F., Compère, C. & Bureau, B. (2009). Chalcogenide Glass Optical Waveguides for Infrared. Biosensing. *Sensors*,Vol. 9, pp.7398-7411,ISSN1424-8220

Anne, M.L., Keirsse, J., Nazabal, V., Hyodo, K., Inoue, S., Pledel, C. B., Lhermite, H., Charrier, J., Yanakata, K., Loreal, O., Person, J. L., Colas, F., Compère, C.& Bureau, B. (2009). Chalcogenide Glass Optical Waveguides for Infrared Biosensing. *Sensors*, Vol. 9, pp.7398-7411, ISSN 1424-8220

Aravinda Narayanan, R., Asokan, S. & Kumar, A. (2001). Influence of chemical disorder on electrical switching in chalcogenide glasses, *Physical Review B*, Vol.63, pp.092203-1-092203-4,ISSN 0163-1829

Augis, J. A.& Bennett, J. E. (1978). Calculation of the Avrami parameters for heterogeneous solid state reactions using a modification of the Kissinger method. *Journal of Thermal Analysis and Calorimetry*,Vol. 13, No.2, pp. 283-292, ISSN 1572-8943

Avrami, M . (1939). Kinetics of Phase Change. I General Theory. *Journal of Chemical Physics*, Vol. 7,pp. 1103-1112,ISSN.1089-769

Avrami, M. (1940). Kinetics of Phase Change. II Transformation-Time Relations for Random Distribution of Nuclei, *Journal of Chemical Physics*, Vol. 8,pp. 212-224, ISSN.1089-769

Bicerono, J. & Ovshinsky, S.R. (1985). Chemical Bond Approach to the Structure of chalcogenide glasses with Reversible Switching Properties, *Journal of Non-Crystalline Solids*, Vol.74, pp. 75-84,ISSN 0022-3093

Bichara, C. & pellegatti, A. (1993). Chain structure of amorphous selenium by tight-binding Mont Carlo simulation. *Physical Review B*, Vol.49, pp.6581-6586, ISSN 1943-2879

Bohmer, R. & Angell, C.A. (1993). Elastic and viscoelastic properties of amorphous and identification of phase transition between ring and chain structures. *Physical Review B*, Vol.48, pp.5857-5864,ISSN 1943-2879

Boo, B. H., Cho, H. & Kang, D. E. (2007). Ab initio and DFT investigation of structures and energies of low-lying isomers of Zn_xSe_x (x = 1–4) clusters. *Journal of Molecular Structure: THEOCHEM*, 806, 77-83, ISSN: 0166-1280

Boolchand, P., Georgiev, D. G. & Goodmana, B. (2001). Discovery of The Intermediate Phase in Chalcogenide Glasses, *Journal of Optoelectronics and Advanced Materials*, Vol.3, pp.703-720,ISSN 1454-4164

Bosch,V., Sánchez, A. J. R., Rojas, F. S. & Ojeda, C. B. (2007). Recent Development in Optical Fiber Biosensors. *Sensors,* Vol. 7, pp.797-859, ISSN 1424-8220

Bowden, B. F.& Harrington, J. A. (2009). Fabrication and characterization of chalcogenide glass for hollow Bragg fibers. *Applied Optics,*Vol. 48, pp. 3050-3054, ISSN 2155-3165

Boycheva, S. & Vassilev, V. (2002). Electrode-limited conductivity of amorphous chalcogenide thin films from the $GeSe_2$-Sb_2Se_3-ZnSe system , *Journal of Optoelectronics and Advanced Materials*, Vol. 4, pp. 33-40, ISSN 1454 - 4164

Caprion, D. & Schober, H. R. (2000). Structure and relaxation in liquid and amorphous selenium. *Physical Review B*, Vol.62, pp.3709-3716, ISSN 1943-2879

Caprion, D. & Schober, H. R. (2002). Influence of the quench rate and the pressure on the glass transition temperature in selenium. *Physical Review B*, Vol.117, pp.2814-2818, ISSN 1943-2879

Carthy, T.M. & Kanatzidis, M.G. (1996). Synthesis in molten alkali metal polythiophosphate fluxes. The new quaternary bismuth and antimony thiophosphates $ABiP_2S_7$ (A = K, Rb), $A_3M(PS_4)_2$ (A = K, Rb, Cs; M = Sb, Bi), $Cs_3Bi_2(PS_4)_3$, and $Na_{0.16}Bi_{1.28}P_2S_6$. *Journal of Alloys and Compounds,* Vol.236, pp.70-85, ISSN 0925- 8388

Ciureanu, P. & Middelhoek, S. (1992). Thin Film Resistive Sensors London, *IOP Publishing*

Cohen, M. H. & Lucovsky, G. (1971). Amorphous and Liquid Semiconductors, North Holland Publishers, Amsterdam.

Dahshan, A., Aly, K.A. & Dessouky, M.T. (2008). Thermal stability and activation energy of some compositions of Ge-Te-Cu chalcogenide system. *Philosophical Magazine,* Vol. 88, pp.2399–2410, ISSN ISSN 1478-6443

Demarco, P. & Pejcic, B. (1999). Electrochemical Impedance Spectroscopy and X-ray Photoelectron Spectroscopy Study of the Response Mechanism of the Chalcogenide Glass Membrane Iron(III) Ion-Selective Electrode in Saline Media, *Analytical Chemistry*, Vol. 72, pp. 669-679, ISSN 0003-2700

Dietzel, A. (1968). Glass Structure and Glass Properties, I, *Glass Science and Technology*, Vol.22,pp. 41-50, ISSN 0017-1085.

Doremus, W. (1969). Semiconductor Effects in Amorphous Solids, North Holland Publishers, Amsterdam

Dudley, J. M. & Taylor, J. R. (2009). Ten years of nonlinear optics in photonic crystal fibre.*Nature Photonics,* Vol. 3, pp. 85-90, ISSN 1749-4885

Echeveria, I., Kolek, P. L., Plazek, D. J. & Simon, S. L. (2003). Enthalpy recovery, creep and creep–recovery measurements during physical aging of amorphous selenium. *Journal of Non-Crystalline Solids*, Vol.324, pp.242-255, ISSN 0022-3093

Efimov, A. M. (1999). Vibrational spectra, related properties, and structure of inorganic glasses. *Journal of Non-Crystalline Solids*. Vol. 253,pp. 95-118, ISSN 0022-3093

Eggleton, B. J., Davies, B. L. & Richardson, K. (2011). Chalcogenide photonics, *Nature Photonics* ,Vol. 5,pp.141–148, ISSN

Eggleton, B.J., Davies, B. L. & Richardson, K. (2011). Chalcogenide photonics, Nature Photonics, Vol. 5, pp.141-148, ISSN 1749-4885

Elliott, G. R., Senthil Murugan, G., Wilkinson, J. S., Zervas, M. N.& Hewak, D. W. (2010) Chalcogenide glass microsphere laser. *Optical Express*,Vol. 18,pp. 26720-26727,ISSN 1094-4087

Emin, D. (2006). Current-driven threshold switching of a small polaron semiconductor to a metastable conductor. *Physical Review B* 74, 035206-10, ISSN 1943-2879

Reid, R. C., Botsaris, G. D., Margolis, G., Kirwan, D.J., Denk, E. G., Ersan, G.S., Tester, J. & Wong, F. (1970). Crystallization—Part I - Transport Phenomena of Nucleation and Crystal Growth. *Industrial & Engineering Chemistry*, Vol. 62, pp 52–67, ISSN 0888-5885

Fayek, S. A., El Sayed, S. M., Mehana, A. & Hamza, A. M. (2001). Glass formation, optical properties and local atomic arrangement of chalcogenide systems GeTe-Cu and GeTe-In, *Journal of Materials Science*, Vol. 36, pp.2061-2066,ISSN 0022-2461

Flaschen, S.S., Pearson, A.D., Northover, W.R. (1959). Low-Melting Inorganic Glasses with High Melt Fluidities Below 400°C. *Journal of the American Ceramic Society*, Vol.42, pp.450, ISSN 1551-2916

Fraser, W. A. & Jerger, J. (1953). New Optical Glasses with Good Transparency in the Infrared. *Journal of Optical Society of America*, Vol. 43, pp.332A, ISSN 1084-7529

Frerichs, R. (1950). "New optical glasses transparent in the Infra-red up to 12 ,u". *Physical Review*, Vol.78, pp.643- 643, ISSN 1050-2947

Frerichs, R. (1953). New optical glasses with good transparency in the infrared. *Optical Society ofAmerica*, Vol.43, pp.1153- 1157, ISSN 1084-7529

Fritzsche, H. (1971). Optical and electrical energy gaps in amorphous semiconductors. *Journal of Non-Crystalline* Solids, Vol.6, pp.49-71,ISSN 0022-3093

Gill, W.D. & Street, G.B. (1973). Drift mobility in amorphous selenium-sulfur alloys. *Journal of Non-Crystalline Solids*, Vol.13, pp.120,ISSN ISSN 0022-3093

Goetzberger, A. & Hebling, C. (2000). Photovoltaic materials, past, present, future. *Solar Energy Materials & Solar Cells*, Vol. 62, pp.1-19, ISSN 0927-0248

Guo, H., Zhai, Y., Haizheng, T., Yueqiu, G. & Zhao, X. (2007). Synthesis and properties of GeS$_2$–Ga$_2$S$_3$–PbI$_2$ chalcohalide glasses, *Materials Research Bulletin*, Vol. 42, pp.1111-1118, ISSN0025-5408.

Hafner, H. C. (1972) Final Technical Report, AFML-TR-72-54

Hamada, S., Sato, T. & Shirai, T. (1968). Glass Transition Temperature and Isothermal Volume Change of Selinium. *Bulletin of the Chemical Society of Japan*, Vol.41, pp.135-139, 1348-0634

Hawes, L. (1963). Sulphur–Selenium and Sulphur–Tellurium Cyclic Interchalcogen Compounds.*Nature*, Vol. 198, pp.1267-1270, ISSN0028-0836

Hegab, N. A. & Atyia, H. E. (2007). Dielectric Studies Of Amorphous $As_{45}Te_{33}Ge_{10}$ Si_{12} Films, *Journal of Ovonic Research*, Vol.3, pp.93-102, ISSN 1584 - 9953

Hilton, A. R. & Brau, M. (1963). New high temperature infrared transmitting glasses. *Infrared Physics*, Vol.3,pp. 69-72, ISSN 1350-4495

Hilton, A. R. (1966). Nonoxide Chalcogenide Glasses as Infrared Optical Materials. Applied Optics, Vol. 5, pp. 1877-1882, ISSN 2155-3165

Hilton, A. R. (2010). Chalcogenide Glasses for Infrared Optics, Chapter-2,*The McGraw-Hill companies*, pp.17-59, ISBN 978-0-07-159698-5

Hilton, A. R. (June 1974). Defense Advanced Research Projects Agency (DARPA), Contract No. N00014-73-C-0367.

Hruby, A. (1972). Evaluation of glass-forming tendency by means of DTA. *Czechoslovak Journal of Physics*, Vol.22, No.11, pp. 1187-1193, ISSN1572-9486

Hsu, J. W.P. (2001). Near-field scanning optical microscopy studies of electronic and photonic materials and devices. *Materials Science and Engineering. R:Reports*, Vol. 33, pp.1-50, ISSN 0927-796X

Ielmini, D. (2008). Threshold switching mechanism by high-field energy gain in the hopping transport of chalcogenide glasses. *Physical   Review B*, Vol.78, pp.035308 -8, ISSN 0163-1829

Imran, M. M. A., Bhandari, D.& Saxena, N. S. (2001). Enthalpy recovery during structural relaxation of $Se_{96}In_4$ chalcogenide glass. *Physica B*,Vol.294,pp.394-401, ISSN 0921-4526

Ionov, R., Nesheva, D. & Arsova, D. (1991). Electrical and electrophotographic properties of CdSe/SeTe and CdSe/Se multilayers, *Journal of Non-Crystalline Solids*. Vol. 137,pp. 1151-1154, ISSN 0022-3093

Ivan, I.& Kikineshi, A. (2002).Stimulated interdiffusion and expansion in amorphous chalcogenide multilayers. *Journal of Optoelectronics   and Advanced Materials*,Vol.4, pp.743 – 746, ISSN 2066-0049

Ivanova, Z.G., Cernoskova, E., Vassilev, V.S. & Boycheva, S.V. (2003). Thermomechanical and structural characterization of $GeSe_2$–Sb_2Se_3–ZnSe glasses, *Materials Letters*, Vol. 57, pp. 1025-1028, ISSN 0167-577X

JablÇonska, A., Burian, A., Burian, A.M., Lecante, P. & Mosset, A., (2001). Modelling studies of amorphous In–Se films. *Journal of Alloys and Compounds*, Vol.328, pp.214-217, ISSN: 0925-8388

Jackson, K. & Srinivas, S. (2002). Modeling the [119]Sn Mössbauer spectra of chalcogenide glasses using density-functional theory calculations. *Physical   Review B*, Vol. 65, pp.214201-8, ISSN 1943-2879

Jain, P. K., *Deepika & Saxena, N. S.*, (2009).Glass transition, thermal stability and glass-forming ability of $Se_{90}In_{10-x}Sb_x$ (x = 0, 2, 4, 6, 8, 10) chalcogenide glasses, *Philosophical Magazine*,Vol. 89,pp. 641-650, ISSN 1478-6443

Johnson, W. A. & Mehl, R. F. (1939). Reaction kinetics in processes of nucleation and growth. *Transactions of the American Institute of Mining, Metallurgical, Engineers* ,Vol.135,pp. 416-458, ISSN 0096-4778

Jones, C. & Hafner, H. (1968) Final Technical Report, Contract No. AF 33 (615)-3963

Joseph, C . (2010). Crystal Growth: Theory and Techniques. Chapter-1
   (*shodhganga.inflibnet.ac.in/bitstream/10603/265/8/08_chapter1.pdf*)

Kawamura,T.,Yamamoto, N., & Nakayama, Y. (1983). Electrophtographic Application of
   Amorphous Semiconductors, Amorphous Semiconductor Technologies and
   Devices, *JARECT Hamakawa, Y. North Holland OHMSHA,*Vol.6, pp.325-336

Khan, M.A.M., Khan, M.W., Husain, M. & Zulfequar, M. (2009). Electrical transport and
   optical properties of Zn doped Bi-Se chalcogenide glasses. *Journal of Alloys and
   Compounds.* Vol.486, pp.876– 880, ISSN 0925- 8388

Kissinger, H. E. (1957).Reaction Kinetics in Differential Thermal Analysis. *Analytical
   Chemistry,* Vol. 29, pp.1702-1706, ISSN1520-6882

Klokishner, S.I., Kulikova, O.V., Kulyuk, L.L., Nateprov, A.A., Nateprov, A.N., Ostrovsky,
   S.M., Palii, A.V., Reu, O.S. & Siminel, A.V. (2008). Concentration effects in the
   photoluminescence spectra of $ZnAl_{2(1-x)}Cr_{2x}S_4$. *Optical Materials,* Vol. 31, pp.284–
   290, ISSN 0925-3467

Kobelke, J. , Kirshhof, J., Scheffer, S. & Schuwuchow, A. (1999). Chalcogenide glass single
   mode fibres – preparation and properties, *Journal of Non-Crystalline Solids,* Vol.256–
   257, pp. 226-231, ISSN 0022-3093

Kohary, K., Burlakov, V. M., Pettifor,   D. G. & Manh, D. N. (2005). Modeling In-Se
   amorphous alloys. *Physical Review B,* Vol.71, pp.184203-1- 184203-7, ISSN 0022-3093

Kokenyesi, S. (2006). Amorphous chalcogenide nano-multilayers: research and
   development. *Journal of Optoelectronics and Advanced Materials,* Vol. 8, pp.2093 –
   2096, ISSN 2066-0049

Kokenyesi, S., Takats, V., Ivan, I., Csik, A., Szabo, I., Beke, D., Nemec, P., Sangunni, K., &
   Shiplyak, M. (2007). Amorphous chalcogenide nano-multilayers: research and
   development. *Acta Physica Debreceniensis,* Vol. XLI, pp. 51-57, ISSN 1789-6088

Kokorina, V. (1996). Glasses for Infrared Optics, *CRC Press, Boca Raton, Florida,* ISBN: 0-8493-
   3785-2

Kumar, S., Singh, K. & Mehta, N. (2010). Calorimetric studies of crystallisation kinetics of
   $Se_{75}Te_{15-x}Cd_{10}In_x$ multi-component chalcogenide glasses using non-isothermal DSC.
   Philosophical Magazine Letters, Vol.90, pp.547–557, ISSN 1362-3036

Lenz, G., Zimmermann, J.K. & Katsufuji, T. (2000). Large kerr effect in bulk Se-based
   chalcogenide glasses, *Optical Letter,*Vol. 25, pp. 254 -256,ISSN 1539-4794

Lezal, D., Pedlikov, J. & Zavadil, J. (2004). Chalcogenide glasses for optical and photonics
   applications. *Chalcogenide Letteres,* Vol. 1, pp.11 – 15, ISSN ISSN 1584-8663

Lopez, C., (2004).Evaluation of the photo-induced structural Mechanisms in chalcogenide
   glass materials, Thesis (Ph.D.) *University of Central Florida,* pp.1-213, Publication
   Number: AAI3163612; ISBN: 9780496978441

Lousteau, J., Furniss, D., Arrand, H.F., Benson, T.M., Sewell, P.& Seddon, A.B. (2008).
   Fabrication of heavy metal fluoride glass, optical planar waveguides by hot-spin
   casting. *Journal of Non-Crystalline Solids,* Vol. 354, pp.3877–3886, ISSN 0022-3093

Maharjan, N.B., Bhandari, D., Saxena, N.S., Paudyal, D.D.& Husain, M. (2000). Kinetic
   Studies of Bulk $Se_{85-x}Te_{15}Sb_x$ Glasses with x = 0, 2, 4, 6, 8 and 10 . *physica status
   solidi (a),* Vol.178, pp.663-670,ISSN 1862-6319

Malek, J., Pustkova, P. & Shanilova, J. (2003). Kinetic Phenomena In Non-Crystalline Materials Studied By Thermal Analysis, *Journal of Thermal Analysis and Calorimetery*,Vol.72, pp.289-297, ISSN 1388-6150

Malek, J., Svoboda, R., Pustkova, P. & Cicmanec, P. (2009). Volume and enthalpy relaxation of a-Se in the glass transition region. *Journal of Non-Crystalline Solids*, Vol.355,pp.264-272, ISSN 0022-3093

Matusita, K., Komatsu, T.& Yokota, R. (1984).Kinetics of non-isothermal crystallization process and activation energy for crystal growth in amorphous materials. *Journal Of Materials Science*, Vol.19, No.1, pp. 291-296, ISSN 1573-4803

Mehta, N. & Kumar, A. (2007). Comparative Analysis Of Calorimetric Studies In $Se_{90}M_{10}$ (M=In, Te, Sb) Chalcogenide Glasses, *Journal of Thermal Analysis and Calorimetery*, Vol.87, pp.345-150,ISSN 1388-6150

Mehta, N., Singh, K. & Kumar, S. (2009).Effect of Sb and Sn additives on the activation energies of glass transition and crystallization in binary $Se_{85}Te_{15}$ alloy. *Phase Transitions*, Vol. 82, pp.43–51, ISSN 0141-1594

Mehta, N., Tiwari, R.S. & Kumar, A. (2006). Glass forming ability and thermal stability of some Se–Sb glassy alloys, *Materials Research Bulletin*, Vol. 41,pp. 1664-1672,ESSN0025-5408

Mersmann, A. (2001). *Crystallization Technology Handbook*, CRC; 2nd ed. ISBN 0-8247-0528-9

Messaddeq, S. H., Tikhomirov, V. K., Messaddeq, Y., Lezal, D.& Siu Li, M. (2001). Light-induced relief gratings and a mechanism of metastable light-induced expansion in chalcogenide glasses. *Physical Review B*, Vol.63, pp.224203-5, ISSN 1943-2879

Micoulaut, M. & Phillips, J. C. (2003). Rings and rigidity transitions in network glasses, *Physical Review B*, Vol. 67, pp.104204 -9, ISSN 1943-2879

Milliron, D. J., Raoux, S., Shelby, R. M. & Sweet, J. J. (2007). Solution-phase deposition and nanopatterning of GeSbSe phase-change materials.*Nature Materials* Vol.6, pp.352, ISSN 1476-4660

Mortensen, J., Legin, A., Ipatov, A., Rudnitskaya, A., Vlasov, Y. & Hjuler, K. (2000). A flow injection system based on chalcogenide glass sensors for the determination of heavy metals. *Analytica Chimica Acta*, Vol. 403, pp.273–277, ISSN 0003-2670

Mott, N.F. & Davis, E.A. (1979). Electronic Processes in the Non-Crystalline Materials, *Oxford University Press, Second Edition*

Mourizina, Y., Yoshinobu, T., Schubert, J., Luth, H., Iwasaki, H.& Schoning, M. J. (2001). Ion-selective light-addressable potentiometric sensor (LAPS) with chalcogenide thin film prepared by pulsed laser deposition. *Sensor and Actuators B*, Vol. 80, pp.136-140, ISSN 0925-4005

Moynihan, C.T. , Easteal, A.J., Wilder, J., Tucker, J. (1974). Dependence of the glass transition temperature on heating and cooling rates. *Journal of Physical Chemistry*, Vol.78,No. 26,pp.2673-2677, ISSN 1520-5207

Narayanan, R. A., Asokan, S. & Kumar, A. (2001). Influence of chemical disorder on electrical switching in chalcogenide glasses. *Physical Review B*, Vol. 63,pp. 092203 -4, ISSN 1943-2879

Natale, C. D., Davide, F., Brunink, J. A.J., Amico, A. D., Vlasov, Y. G., Legin, A. V. & Rudnitskaya, A. M. (1996). Multicomponent analysis of heavy metal cations and inorganic anions in liquids by a non-selective chalcogenide glass sensor array. *Sensors and Actuators B* Vol.34, pp.539-542, ISSN 0925-4005

Naumis, G. G. (2000). Contribution of floppy modes to the heat capacity jump and fragility in chalcogenide glasses. *Physical Review B-Rapid Communications*, Vol. 61, pp.9205-9208, ISSN 1089-4896

Nesheva, D., Arsova, D.& Vateva, E. (1997). Electrophotographic photoreceptors including selenium-based multilayers. *Semiconductor Science and Technology*, Vol.12, pp.595-599, ISSN 1361-6641

Nielsen, S. (1962). Note on the preparation and properties of glasses containing germanium disulphide. *Infrared Physics*, Vol. 2, pp.117-119,ISSN 1350-4495

Othman, A.A., Aly, K.A. & Abousehly, A.M. (2006). Crystallization kinetics in new $Sb_{14}As_{29}Se_{52}Te_5$ amorphous glass, *Solid State Communications*,Vol.138, pp.184-189,ISSN 0038- 1098

Ovshinsky, S.R.(1994). An History of Phase Change Technology, *Memories Optics Systems*,Vol. 127, pp. 65, ISSN

Ozawa, T. (1970). Kinetic analysis of derivative curves in thermal analysis. *Journal of Thermal Analysis and Calorimetry*,Vol.2, No.3, pp. 301-324, ISSN 1572-8943

Part 1, Hilton, A. R., Jones, C .E. & Brau, M., Part 2, Hilton, A. R. & Jones, C. E., Part 3, Hilton, A. R., Jones, C. E., Dobrott, R. D., Klein, H. M., Bryant, A. M. & George, T. D. (1966). Non-oxide I V A-V A-VIA chalcogenide glasses. *Physics and Chemistry of Glasses*, Vol. 7, pp.105-126, ISSN 0031-9090

Patial, B. S., Thakur, N. & Tripathi, S.K. (2011). On the crystallization kinetics of In additive Se–Te chalcogenide glasses, *Thermochimica Acta*,Vol. 513, pp.1-8, ISSN 0040-6031

Patterson, R. J. (1966)15th National Infrared Information Symposium (IRIS) at Ft Monmouth N.J., (1967) U.S. Patent 3,360,649.

Pearson, A. D. (1962). Electrochemical Society Meeting, *Los Angeles, California-121st Meeting*

Pelusi, M., Luan, F., Vo, T. D., Lamont, M. R. E., Madden, S. J., Bulla, D. A., Choi, D. Y., Davies, B. L.& Eggleton, B. J. (2009). Photonic-chip-based radio-frequency spectrum analyser with terahertz bandwidth. *Nature Photonics*, Vol. 3, pp.139-143, ISSN 1749-4885

Pena, E.Y., Mejia, M., Reyes, J.A., Valladares, R.M., Alvarez, F. & Valladares, A.A. (2004). Amorphous alloys of $C_{0.5}Si_{0.5}$, $Si_{0.5}Ge_{0.5}$ and $In_{0.5}Se_{0.5}$:atomic topology, *Journal of Non-Crystalline Solids*, Vol.338, pp.258-261, ISSN 0022-3093

Peng, H. & Liu, Z. (2010). Organic charge-transfer complexes for STM-based thermochemical-hole-burning memory. *Coordination Chemistry Review*. Vol.254, pp.1151–1168, ISSN 0010-8545

Phillips, J. C. (2006). Ideally glassy hydrogen-bonded networks. *Physical Review B*, Vol. 73, pp.024210 -10, ISSN 1943-2879

Prashanth, S.B. B. & Asokan, S. (2008). Composition dependent electrical switching in $Ge_xSe_{35-x}Te_{65}$ ($18 \leq x \leq 25$) glasses – the influence of network rigidity and thermal properties, *Solid State Communications*, Vol. 147, pp.452-456,ISSN 0038- 1098

Pungor, E. (1997) . Ion-selective electrodes – istory and conclusions, *Journal of Analytical Chemistry*, Vol. 357, pp. 184-188, ISSN 1061-9348

Quiroga, I., Corredor, C., Bellido, F., Vazquez, J., Villares, P. & Garay, R. J. (1996). Infrared studies of Ge-Sb-Se glassy semiconductor, *Journal of Non- Crystalline Solids* Vol.196, pp.183-186,ISSN0022-3093

Rau, C., Armand, P., Pradel, A., Varsamis, C. P. E., Kamitsos, E. I., Granier, D., Ibanez, A. & Philippot, E. (2001). Mixed cation effect in chalcogenide glasses $Rb_2S-Ag_2S-GeS_2$. *Physical Review B*, Vol. 63, pp.184204-9, ISSN 1943-2879

Robert, E.J., Kasap, S.O., Rowlands, J. & Polischuk, B. (1998). Metallic electric contacts to stabilized amorphous selenium for use in X-ray image detectors, *Journal of Non-Crystalline Solids*,Vol. 2,pp. 227-240, 1359-1362, ISSN 0022-3093

Saad, M.& Poulin, M. (1987). Glass Forming Ability Criterion.*Materials Science Forum*,Vol. 19–20,pp. 11-18,ISSN 1662-9752.

Saffarini, G. (2002) The effect of compositional variations on the glass-transition and crystallisation temperatures in Ge-Se-In glasses. *Applied Physics A*,Vol.74, No.2, 283-285,ISSN 1432-0630

Saleh, Z.M., Williams, G.A.& Taylor, P.C. (1993). Nuclear-magnetic-resonance relaxation in glassy Cu-As-Se and Cu-As-S. *Physical Review B*, Vol. 47, pp.4990-5001, ISSN 1943-2879

Salmon, P. S. & Xin, S. (2002). The effect of covalent versus ionic bonding in chalcohalide glasses:$(CuI)_{0.6}(Sb_2Se_3)_{0.4}$. *Physical Review B*, Vol. 65, pp.064202-4, ISSN 1943-2879

Samson, Z. L., Yen, S. C., MacDonald, K. F., Knight, K., Li , S., Hewak, D. W., Tsai, D.P. & Zheludev, N. I. (2010). Chalcogenide glasses in active plasmonics. *Physica Status Solidi RRL*, Vol.4,pp. 274–276, ISSN 1862-6270

Sanchez, E. A., Waldmann, M.& Arnold, C. B. (2011). Chalcogenide glass microlenses by inkjet printing," *Applied Optics*, Vol. 50,pp. 1974-1978 ISSN 2155-3165

Sanchez-Jimenez, P.E. et al., (2009).Combined kinetic analysis of thermal degradation of polymeric materials under any thermal pathway. *Polymer Degradation and Stability*, Vol. 94, pp. 2079-2085,ISSN 0141-3910

Sargent, E. H. (2009). Infrared photovoltaics made by solution processing.*Nature Photonics*,Vol. 3,pp. 325-331, ISSN 1749-4885

Sarrach, J., de Neufville, J.P. & Haworth, W.L. (1976).Studies of amorphous Ge☐Se☐Te alloys (I): Preparation and calorimetric observations. *Journal of Non-Crystalline Solids*, Vol. 22, PP.245, ISSN ISSN 0022-3093

Savage, J. A. & Nielsen, S. (1964). Preparation of glasses transmitting in the infra-red between 8 and 15 microns. *Physics and Chemistry of Glasses*, Vol.5, pp.82-86, ISSN 0031-9090

Savage, J. A.& Nielsen, S. (1966). The infra-red transmission of telluride glasses. *Physics and Chemistry of Glasses*, Vol.7,pp. 56-59, ISSN 0031-9090

Sbirrazzuoli, N. (1999). Isothermal and Non-Isothermal Kinetics When Mechanistic Information Available.*Journal of Thermal Analysis and Calorimetry*,Vol.56, No.2, pp. 783-792, ISSN 1572-8943

Schubert, J., Schoning, M. J., Mourzina, Y.G., Legin, A.V., Vlasov, Y.G., Zander, W.& Luth, H. (2001). Multicomponent thin films for electrochemical sensor applications prepared by pulsed laser deposition. *Sensor and Actuators B*,Vol. 76, pp.327-330, ISSN 0925-4005

Sellack, C. S. (1870). An early, but only qualitative, observation of good transmission of A_2S_3 in the infrared. *Ann. Physik,* Vol. 215, pp. 182–187, ISSN 1521-3889.

Selvaraju, V.C., Asokan, S. & Srinivasan, V. (2003). Electrical switching studies on $As_{40}Te_{60-x}Se_x$ and $As_{35}Te_{65-x}Se_x$ glasse. *Applied Physics A,* Vol. 77, No.1, pp. 149-153, ISSN 1432-0630

Shportko, K., Kremers, S., Woda, M., Lencer, D., Robertson, J. & Wuttig, M. (2008). Resonant bonding in crystalline phase-change materials.*Nature Materials,*Vol. 7,pp. 653-658, ISSN 1476-4660

Singh, A. K. & Singh, K.(2010). Observation of Meyer Neldel rule and crystallization rate constant stability for $Se_{93-x}Zn_2Te_5In$ x chalcogenide glasses. *The European Physical Journal - Applied Physics,*Vol.51, pp.30301 (5pp) ISSN 1286-004

Singh, A. K.& Singh, K. (2009). Crystallization kinetics and thermal stability of $Se_{98-x}Zn_2In_x$ chalcogenide glasses, *Philosophical Magazine,* Vol. 89, pp.1457-1472, ISSN 1478-6443

Singh, A.K & Singh, K. (2011). Localized structural growth and kinetics of $Se_{98-x}Zn_2In_x$ ($0 \leq x \leq 10$) amorphous alloys. *Physica Scripta,*Vol.83, No.2, pp. 025605 (6pp),ISSN1402-4896

Singh, A.K, Mehta,N.&Singh, K. (2009). Correlation between Meyer–Neldel rule and phase separation in $Se_{98-x}Zn_2In_x$ chalcogenide glasses.*Current Applied Physics,* Vol.9, pp.807-811,ISSN1567-1739

Singh, A.K, Mehta,N.&Singh, K. (2010). Effect of indium additive on glass-forming ability and thermal stability of Se-Zn-Te chalcogenide glasses. *Philosophical Magazine Letters,* Vol.90,pp.201-208,ISSN 1362-3036

Singh, A.K. (2011). Effect Of Indium Additive on Heat Capacities Of Se-Zn-Te Multicomponent Chalcogenide Glasses.*Chalcogenide Letters,*Vol.8, No.2,pp.123-128, ISSN 1584-8663

Singh, A.K. (2011). Effect of indium additive on the heat capacity of Se-Zn chalcogenide glasses, *The European Physical Journal Applied Physics,* Vol. 55, pp. 11103-1-11103-4, ISSN 1286-0042

Snopatin, G. E., Shiryaev, V. S., Plotnichenko, V. G., Dianov, E. M. & Churbanov, M. F. (2009). High-purity chalcogenide glasses for fiber optics. *Inorganic Materials,* Vol.45, pp.1439–1460, ISSN 0020-1685

Soltan, A.S., Abu EL-Oyoun, M., Abu-Sehly, A.A. & Abdel-Latief, A.Y. (2003). Thermal annealing dependence of the structural, optical and electrical properties of selenium –tellerium films, *Materials Chemistry and Physics,* Vol. 82,pp. 101-106,ISSN 0254-0584

Song, S.M., Choi, S.Y. & Yong-Keun, L. (1997). Crystallization property effects in $Ge_{30}Se_{60}Te_{10}$ glass, *Journal of Non-Crystalline Solids,* Vol.217,pp.79-82, ISSN 0022-3093

Stocker, H. J. (1969). Bulk and thin film switching and memory effects in semiconducting chalcogenide glasses .*Applied Physics Letters,* Vol.15, pp. 55-57, ISSN1432-0630

Suri, N., Bindra, K.S., Kumar, P., Kamboj, M.S.& Thangaraj, R. (2006). Thermal Investigations In Bulk $Se_{80-x}Te_{20}Bi_x$ Chalcogenide Glass. *Journal of Ovonic Research,* Vol. 2, No. 6, pp. 111 – 118,ISSN 1584 - 9953

Swanson, S.E. (1977). Relation of nucleation and crystal-growth rate to the development of granitic textures. *American Mineralogist,* Vol.62, pp. 966-978, ISSN 0003-004X

Taeed, V. G., Baker, N. J., Fu, L., Finsterbusch, K., Lamont, M. R.E., Moss, D. J., Nguyen, H. C., Eggleton, B. J., Yong Choi, D., Madden, S. & Davies, B. L. (2007). Ultrafast all-optical chalcogenide glass photonic circuits. *Optics Express*, Vol.15, pp.9205- 9221, ISSN 1094-4087

Tanaka, K. (2003). Nanostructured chalcogenide glasses. *Journal of Non-Crystalline Solids*,Vol. 326-327,pp. 21-28, ISSN 0022-3093

Thingamajig, B., Ganesan, R., Asha Bhat, N., Sangunni, K. S. & Gopal, E. S. R. (2000). Determination of thermal diffusion length in bismuth doped chalcogenide glasses, by photoacoustic technique. *Journal of Optoelectronics and Advanced Materials*, Vol. 2, pp.91-94, ISSN 1454 – 4164

Tikhomirov, V.K., Furniss, D., Seddon, A.B., Savage, J.A., Mason, P.D.,Orchard, D.A. & Lewis, K.L. (2004). Glass formation in the Te-enriched part of the quaternary Ge–As–Se–Te system and its implication for mid-infrared optical fibres. *Infrared Physics & Technology*, Vol.45, pp.115-123, SN: 1350-4495

Tonchev, D. & Kasap, S.O. (1999). Thermal properties of Sb_xSe_{100-x} glasses studied by modulated temperature di.erential scanning calorimetry, *Journal of Non-Crystalline Solids*,Vol. 248,pp. 28-36,ISSN 0022-3093

Towers, H. and Chipman, J. (1957). Diffusion of calcium and silicon in a lime-alumina-silica slag. *Transactions of the American Institute of Mining, Metallurgical, Engineers*, Vol. 209, pp.769-773, ISSN 0096-4778

Troles, J. , Niu, Y., Arfuso, D.C. , Smektala, F. , Brilland, L., Nazabal, V., Moizan, V., Desevedavy, F. & Houizot, P. (2008).Synthesis and characterization of chalcogenide glasses from the system Ga–Ge–Sb–S and preparation of a single-mode fiber at 1.55 µm , *Materials Research Bulletin*,Vol. 43, pp. 976-982

Turek, M., Heiden, W., Riesen, A., Chhabda, T.A., Schubert, J., Zander, W., Kruger, P., Keusgen, M. & Schoning, M.J. (2009). Artificial intelligence/fuzzy logic method for analysis of combined signals from heavy metal chemical sensors. *Electrochimica Acta*, Vol. 54, pp.6082–6088, ISSN 0013-4686

Usuki, T., Uemura, O., Konno, S., Kameda, Y. & Sakurai, M. (2001). Structural and physical properties of Ag-As-Te glasses, *Journal of Non-Crystalline Solids*, Vol. 293, pp.799-805, ISSN 0022-3093

Vassilev, V. (2006). Multicomponent Cd (Zn)-containing Ge(As)-chalcogenide glasses. *Journal of the University of Chemical Technology and Metallurgy*, Vol.41, pp.257-276,ISSN 1311-7629

Vassilev, V., Parvanov, S., Vasileva, T. H., Aljihmani, L., Vachkov, V. & Evtimova, T. V. (2007). Glass formation in the As_2Te_3–As_2Se_3–SnTe system, *Materials Letters*, Vol.61,pp. 3676-3578,ISSN 0167-577X

Vassilev, V., Parvanov, S., Vasileva, T. H., Parvanova, V. & Ranova, D. (2007). Glass formation in the As–Te–Sb system, *Materials Chemistry and Physics*,Vol.105, pp.53-57, ISSN 0254-0584

Vassilev, V., Tomova, K., Parvanova, V. & Parvanov, S. (2007). New chalcogenide glasses in the $GeSe_2$–Sb_2Se_3–PbSe system. *Materials Chemistry and Physics*, Vol. 103, pp.312-317, ISSN0254-0584

Vassilev, V., Tomova, K., Parvanova, V.& Boycheva, S. (2009). Glass-formation in the GeSe$_2$–Sb$_2$Se$_3$–SnSe system. *Journal of Alloys and Compounds*,Vol. 485, pp.569-572, ISSN 0925-8388

Vassilev, V.S. & Boycheva, S.V. (2005). Chemical sensors with chalcogenide glassy membranes. *Talanta*, Vol. 67,pp. 20–27, ISSN 0039-9140

Vassilev, V.S., Hadjinikolova, S.H. & Boycheva, S.V. (2005). Zn(II)-ion-selective electrodes based on GeSe$_2$–Sb$_2$Se$_3$–ZnSe glasses , *Sensors   Actuators   B* , Vol. 106, pp. 401-406, ISSN 0925-4005

Velinov, T., Gateshki, M., Arsova, D. & Vateva, E. (1997). Thermal diffusivity of Ge-As-Se(S) glasses, *Physical Review B*, Vol.55, pp.11014-4, ISSN 1943-2879

Volmer, M. and Weber, A. (1925). *Z. Phys. Chem.*, Vol.119, pp.227, ISSN Tammann, G. (1925). The States of Aggregation. *Van Nostrand, New York.*

Wachter, J.B., Chrissafis, K., Petkov, V., Malliakas, C.D., Bilc, D., Kyratsia, Th., Paraskevopoulos, K.M., Mahanti, S.D., Torbrugge, T., Eckert, H. & Kanatzidis, M.G. (2007). Local structure and influence of bonding on the phase-change behavior of the chalcogenide compounds K$_{1-x}$Rb$_x$Sb$_5$S$_8$. *Journal of Solid State Chemistry*, Vol. 180, pp.420–431, ISSN 0022-4596

Wagner, T., Kasap, S.O., Vlcek, M., Sklenar, A. & Stronski, A. (1998). The structure of As$_x$ S$_{100-x}$ glasses studied by temperature-modulated differential scanning calorimetry and Raman spectroscopy, *Journal of Non-Crystalline Solids*,Vol. 2227,pp. 752-756,,ISSN 0022-3093

Wang, R. P., Zha, C. J., Rode, A. V. Madden, S. J. & Davies, B. L. (2007). Thermal characterization of Ge–As–Se glasses by differential scanning calorimetry, *Journal of Materials Science: Materials in Electronics*, Vol. 18, pp.419-422,ISSN 0957-4522

Warghese, G. (2010). Crystal Growth. Chapter -1 (*shodhganga.inflibnet.ac.in/bitstream/10603/.../08_chapter%201.pdf*).

Wibowo, R. A., Kim, W. S., Lee, E. S., Munir, B. & Kim, K. H. (2007). Single step preparation of quaternary Cu$_2$ZnSnSe$_4$ thin films by RF magnetron sputtering from binary chalcogenide targets. *Journal of Physics and Chemistry of Solids*, Vol. 68, pp.1908-1913, ISSN: 0022-3697

Wuttig, M .& Yamada, N. (2007). Phase-change materials for rewriteable data storage.*Nature Materials*,Vol. 6, pp.824-832, ISSN 1476-4660

Xu, Y., Zhang, Q., Wang, W., Zeng, H., Xu, L. & Chen, G. (2008). Large optical Kerr effect in bulk GeSe$_2$–In$_2$Se$_3$–CsI chalcohalide glasses , *Chemical Physics Letters*, Vol.462, pp. 69-71, ISSN 0009-2614

Yang, C.Y., Paesler, M.A. &Sayers, D.E. (1989).Chemical order in the glassy As$_x$S$_{1-x}$ system: An x-ray-absorption spectroscopy study, *Physical Review B*, Vol.39, pp.10342-11, ISSN 1943-2879

Zhang, G., Gu, D., Jiang, X., Chen, Q. & Gan, F. (2005). Femtosecond laser pulse irradiation of Sb-rich AgInSbTe films: Scanning electron microscopy and atomic force microscopy investigations, *Applied Physics A Materials Science & Processing*, Vol. 80, pp. 1039-1043, ISSN 0947-8396

Zhang, X. H., Ma, H., & Lucas, J. (2004). Evaluation of glass fibers from the Ga–Ge–Sb–Se system for infrared applications, *Optical Materials*, Vol. 25, pp.85-89, ISSN 0925-3467

4

Recrystallization of Active Pharmaceutical Ingredients

Nicole Stieger and Wilna Liebenberg
North-West University, Unit for Drug Research & Development
South Africa

1. Introduction

Recrystallization can be described simply as a process whereby a crystalline form of a compound may be obtained from other solid-state forms, being themselves crystalline or amorphous, of the same substance. Recrystallization is the process most often employed for the intermediate separation and last-step purification of solid active pharmaceutical ingredients (APIs) (Shekunov & York, 2000; Tiwary, 2006). Chemical purity is however not the only property of a pharmaceutical active that will affect its performance. Crystal structure (Table 1), crystal habit (Table 2) and particle size all play a part. Polymorphism (different crystal structures of the same substance) affects the physico-chemical properties and stability of an API, whereas crystal habit and particle size mostly affect various indices impacting on dosage form production and performance: particle orientation; flowability; packing and density; surface area; aggregation; compaction; suspension stability; and dissolution (Blagden *et al.*, 2007; Doherty & York, 1988; Tiwary, 2006).

System	Relationship between lattice parameters
Cubic	$a = b = c$ $\alpha = \beta = \gamma = 90°$
Tetragonal	$a = b \neq c$ $\alpha = \beta = \gamma = 90°$
Orthorhombic	$a \neq b \neq c$ $\alpha = \beta = \gamma = 90°$
Rhombohedral (or trigonal)	$a \neq b \neq c$ $\alpha = \beta = \gamma \neq 90°$
Hexagonal	$a = b \neq c$ $\alpha = \beta = 90°, \gamma = 120°$
Monoclinic	$a \neq b \neq c$ $\alpha = \gamma = 90°, \beta \neq 90°$
Triclinic	$a \neq b \neq c$ $\alpha \neq \beta \neq \gamma \neq 90°$

Table 1. Crystal systems and lattice parameters. (Adapted, with permission of Informa Healthcare, from Rodríguez-Homedo *et al.*, 2006.)

Crystal morphology	Description
Acicular	Slender, needle-like crystal
Aggregate	Mass of adhered crystals
Blade	Long, thin, flat crystal
Dendritic	Tiny crystallites forming a tree-like pattern
Equant/cubic	Crystal with similar length, width and thickness
Fiber	Long, thin needle; longer than acicular
Flake/lath	Thin, flat crystal similar in width and length
Plate/platy	Flat crystals with similar width and length but thicker than a flake
Prismatic/bipyramid	Hexagonal crystals with faces parallel to the growth axis; width and thickness greater than acicular and shorter in length
Rod	Cylindrical crystals elongated along one axis
Rosette/spherulite	Sphere composed of needles or rods radiating from a common center
Tablet/tabular	Flat crystal with similar width and length but thicker than a plate

Table 2. Descriptions of common pharmaceutical crystal morphologies. (Adapted, with permission of Informa Healthcare, from Rodríguez-Homedo et al., 2006.)

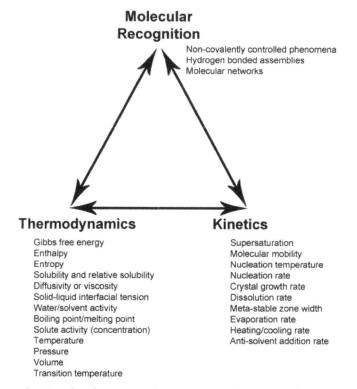

Molecular Recognition

Non-covalently controlled phenomena
Hydrogen bonded assemblies
Molecular networks

Thermodynamics

Gibbs free energy
Enthalpy
Entropy
Solubility and relative solubility
Diffusivity or viscosity
Solid-liquid interfacial tension
Water/solvent activity
Boiling point/melting point
Solute activity (concentration)
Temperature
Pressure
Volume
Transition temperature

Kinetics

Supersaturation
Molecular mobility
Nucleation temperature
Nucleation rate
Crystal growth rate
Dissolution rate
Meta-stable zone width
Evaporation rate
Heating/cooling rate
Anti-solvent addition rate

Fig. 1. Diagram showing the phenomena that govern solid-state transformations. Mechanical, thermal and chemical (solvents, additives, impurities, relative humidity) stresses affect the competition (or reinforcement) among these processes. (Adapted, with permission of Informa Healthcare, from Rodríguez-Homedo *et al.*, 2006.)

The number of solid-state forms that an API could exist in relies on the range of non-covalent interactions and molecular assemblies, the order range, and the balance between entropy and enthalpy that defines the free energy states and processes (Figure 1). When an API exists as more than one solid-state form, thermodynamics control their relative stability and the conditions and direction in which a transformation can occur, whilst kinetics determine how long it will take for a transformation to reach equilibrium. Thermodynamics establish the stability domains of the solid-states, but once a metastable domain is encountered, the kinetic pathways will determine which form will be created and for how long it will survive (Rodríguez-Spong et al., 2004).

Small changes in recrystallization procedure can influence the crystallization process and may lead to changes in API crystal structure, crystal habit and particle size with subsequent variability of raw material characteristics and dosage form performance. Herein lies the challenge and the opportunity: manufacturers of pharmaceutical actives go to great lengths to ensure production uniformity, but a pharmaceutical researcher might choose to alter recrystallization conditions to manipulate API characteristics.

Crystallization is generally thought of as the evolution from solution or melt of the crystalline state (Blagden et al., 2007), but a broader definition includes precipitation and solid-state transitions (Shekunov & York, 2000). Crystallization methods can be solvent or non-solvent based and the varied reaction conditions generate different crystal forms (Banga et al., 2004). Non-solvent based methods include sublimation, thermal treatment, desolvation of solvates, grinding and crystallization of a melt (Guillory, 1999). Applied in polymorph screening studies, traditional crystallization approaches most often will not yield all possible polymorphs of a given API. Therefore, there is a continuous search for innovative methods of manipulating the crystallization process (Rouhi, 2003).

2. Solvent based recrystallization

In solution, crystallization is the creation of a crystalline phase by a process (Figure 2) initiated by molecular aggregation, leading to the formation of nuclei (the smallest possible units with defined crystal lattice) and ultimately crystal growth (Banga et al., 2004).

2.1 Parameters that influence nucleation and crystal growth

The sections that follow will address factors that influence solvent recrystallization in general and will not venture into the specifics of specialized solvent-based production methods like spray-drying, freeze-drying, etc. The reader should keep in mind that although different parameters are discussed under separate headings, they are interactive and not independent of each other. For example: changing the composition of a solvent system will change the level of saturation; changes in temperature alter viscosity and level of saturation; agitation may increase temperature; etc.

Each of the parameters discussed can influence polymorphism, and polymorphism in turn affects crystal habit. It is also well-known that each polymorph of an API can exhibit multiple crystal habits, therefore a parameter might influence crystal habit without changing the internal structure of the crystals produced (Tiwary, 2006).

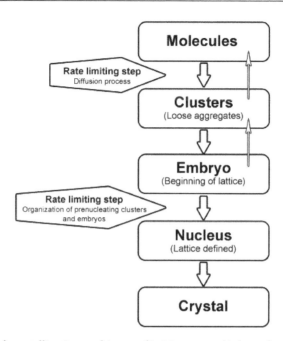

Fig. 2. The course of crystallization and its rate-limiting steps. (Adapted, with permission of Touch Briefings, from Banga *et al.*, 2004.)

2.1.1 Concentration and temperature

The difference in chemical potential between the crystallization solution and the solid phase is the fundamental driving force for crystallization, but it is more convenient to express the driving force in terms of supersaturation (Fujiwara *et al.*, 2005). Supersaturation (the difference between solution concentration and saturation concentration at a specific temperature) leads to the creation of metastable (far from equilibrium) liquid states and crystallization provides a means to reduce the free energy of the system to the most stable state (equilibrium) (Rodríguez-Homedo *et al.*, 2006). The kinetics of nucleation and crystal growth are strongly dependent on supersaturation (Braatz, 2002; Togkalidou *et al.*, 2002). An increase in the degree of supersaturation of a solution leads to a reduction in the size of crystals produced. At high supersaturation nucleation is more rapid than growth, resulting in the precipitation of fine particles (Carstensen *et al.*, 1993; Tiwary, 2006). It is therefore important to control the degree of supersaturation during crystallization, because the size, shape and solid-state form of the crystals produced are all influenced by supersaturation (Fujiwara *et al.*, 2005).

Supersaturation is typically achieved through processes that either increase the solute concentration (evaporation or dissolution of a metastable solid phase with subsequent transformation to the more stable, but less soluble form) or decrease the solubility of the solute (cooling, addition of an antisolvent, pH change or the addition of ions that participate in precipitation of the solute) or through a combination of these strategies (Fujiwara *et al.*, 2005; Rodríguez-Homedo *et al.*, 2006).

The main effect of temperature on crystallization from solution is secondary to its influence on the solubility of the solute, and subsequently the degree of saturation of the solution. Although molecular recognition is required for the formation of molecular clusters, and this is dependent on molecular mobility and collision rates, both of which increase at higher temperatures, the molecular mobility in liquids is too high to be the rate-limiting factor in nucleation and crystal growth (Rodríguez-Spong et al., 2004).

The relationship between nucleation rate and supersaturation is well known and Figure 3 illustrates that the number of nuclei generated by a high rate of cooling increases exponentially with increasing supersaturation. A high number of nuclei at the outset limits their growth potential (Tung et al., 2009).

In cases where the crystallization of a metastable polymorph precedes in-solvent transformation to a more stable polymorph as described by Ostwald's Rule of Stages (Boistelle & Astier, 1988; Ostwald, 1897), the solvent-mediated transformation can be affected by the temperature of the crystallization medium (Stieger et al., 2009), because the process is thermally activated (Beckman, 2000).

Temperature cycling or oscillation is sometimes used to accelerate the effect of post-crystallization Ostwald Ripening in slurries containing just one polymorph of an API. Small particles and rough edges of larger particles dissolve faster during heating periods, followed by their recrystallization onto the existing crystals during cooling. The overall effect is more uniformly shaped particles and a narrower particle size distribution range (Kim et al., 2003; Mullin, 2001).

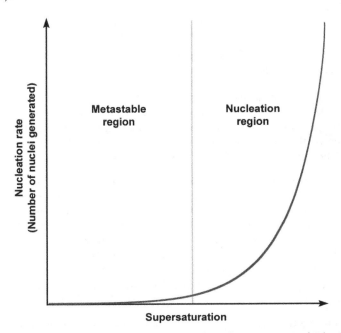

Fig. 3. Nucleation versus supersaturation (Adapted, with permission of John Wiley and Sons, from Tung et al., 2009).

2.1.2 Interfaces and surfaces

Nucleation can be either homogeneous or surface catalyzed (Figure 4). Homogeneous nucleation seldom occurs in volumes greater than 100 µl, because solutions contain random impurities that may induce nucleation. Surface catalyzed nucleation can be promoted by surfaces of the crystallizing solute (secondary nucleation) or a surface/interface of different composition than the solute may induce nucleation (heterogeneous nucleation) by decreasing the energy barrier for the formation of nuclei (Rodríguez-Homedo *et al.*, 2006).

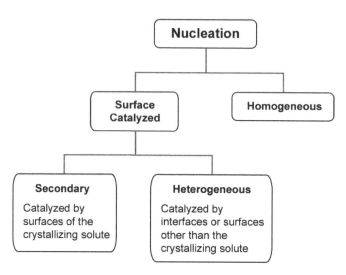

Fig. 4. Mechanisms for crystal nucleation. (Adapted, with permission of Informa Healthcare, from Rodríguez-Homedo *et al.*, 2006.)

Surfaces promoting heterogeneous nucleation may be introduced into the crystallization solution intentionally (as a means of controlling crystal form of the product) (Rodríguez-Spong *et al.*, 2004) or unintentionally (dust and other impurities), or they may be an unavoidable part of the process (crystallization vessel, vessel-solution interface and the solution-air interface) (Florence & Attwood, 2006; Kuzmenko *et al.*, 2001).

The intentional introduction, into crystallization solutions, of surfaces that catalyze nucleation is known as "seeding". Usually, seeding is performed by introducing crystals of the solute that have the preferred crystal structure one wishes to obtain. Seeding can also be performed using isomorphous substances that differ from the solute (Florence & Attwood, 2006). Seeding techniques can be applied to initiate crystallization; to control particle size – usually when larger crystals with uniform size distribution are required; to avoid encrustation through spontaneous nucleation; to control polymorphic form; and to obtain crystals of high purity, high perfection, desired orientation and sufficient size for crystal structure determination by X-ray diffraction (Beckman, 2000).

When the concentration of an API in the crystallization medium is increased past its solubility curve (Figure 5), whether by cooling or by evaporation of solvent, nucleation does not immediately occur. The solution has to reach a certain concentration-temperature point where spontaneous nucleation occurs – this is the border of the metastable zone. For seeding to be successful in producing just the required product, the concentration and temperature of the crystallization medium must be strictly controlled within this zone. If the solubility line is crossed to the left, the seed crystals will dissolve and if the metastable zone limit is crossed to the right, spontaneous nucleation will take place and could result in unwanted crystal forms (Beckman, 2000; Fujiwara *et al.*, 2005).

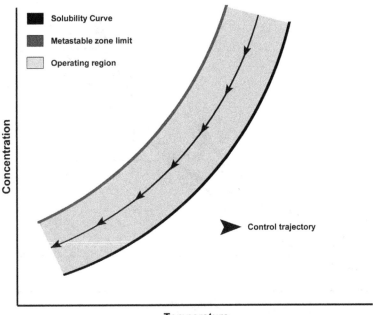

Temperature

Fig. 5. The operating region of seeded industrial batch crystallization is the metastable zone, which is bound by the solubility curve and the metastable limit for the specific API. The concentration-temperature profile (control trajectory) to be used lies within the metastable zone (operating region). (Adapted, with permission of Elsevier, from Fujiwara *et al.*, 2005.)

In evaporative crystallization, crystals are sometimes observed to form preferentially near the surface of the solution, due to a higher local concentration of solute. The meniscus of a solution can also have geometry favoring higher evaporation rates, with crystals then forming at the contact line with the crystallization vessel (Capes & Cameron, 2007).

2.1.3 Solvent

Dependent on the conditions employed, the crystallization of API polymorphs from a solvent may be under kinetic or thermodynamic control. When crystallization of a dimorphic compound (Figure 6) is conducted sufficiently above or below the transition point, in an area defined by the solubility curves of the two polymorphs, the solvent used is immaterial provided that the API solubility is adequate to allow the prescribed concentrations to be reached. Irrespective of the kinetics, the outcome is under total thermodynamic control. If crystallization takes place outside the area described above (B2, C1, D1, E1 and E2), the choice of solvent may or may not be critical. This will be determined by the relative kinetics of formation, growth, and transformation of the two polymorphs in the various solvents (Threlfall, 2000).

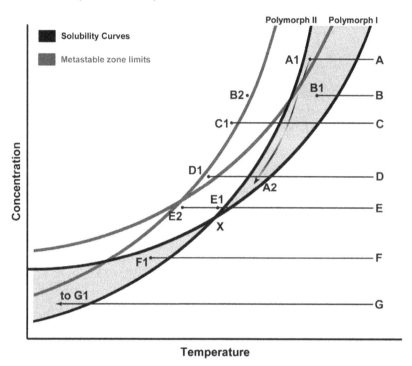

Temperature

Fig. 6. Polymorphic system of two enantiomorphically related polymorphs I and II. (Transition point X at T_X; A-G, initial state of hot, undersaturated solutions; A1-G1, B2 and E2, state of solution at point of initial crystallization. If B is seeded at B1, it behaves as A1.) (Adapted, with permission of the American Chemical Society, from Threlfall, 2000.)

Recrystallization from solvents often leads to the isolation of solvates, in fact, API solvates are very common (Griesser, 2006). Although they have a recognized potential to improve dissolution kinetics (Brittain & Grant, 1999; Haleblian, 1975; Tros de Ilarduya *et al.*, 1997), they are rarely selected for further development or dosage form formulation – the only exceptions being hydrates. The main reasons for the rarity of marketed API solvates are their solid-state metastability and the relative toxicity of any included solvent (Douillet *et al.*, 2011). It goes without saying that, in the production of solvates, the solvent/s used for recrystallization will be the one/s that could potentially be included in the crystal lattice.

The nature of the crystallization solvent can affect crystal habit, regardless of change in polymorphism (Stieger *et al.*, 2010a). The interaction of the solvent at the different crystal-solution interfaces may lead to altered roundness of growing crystal faces and/or edges, changes in crystal growth kinetics, and enhancement or inhibition of crystal growth at certain faces (Figure 7), thereby changing the crystal habit (Stoica *et al.*, 2004; Tiwary, 2006).

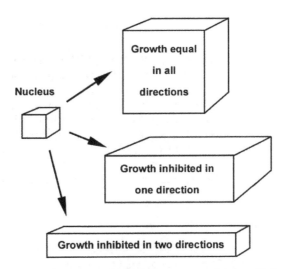

Fig. 7. Crystal habits arising from growth inhibition at crystal faces. (Adapted, with permission of Informa Healthcare, from Carstensen *et al.*, 1993.)

2.1.4 Agitation, mixing and stirring

Mixing of crystallization solutions and crystal slurries is often necessary, especially in industry, to ensure homogeneity (heat transfer, dispersion of additives, uniformity of crystal suspension, avoidance of settling, etc.). Its effects on nucleation, both primary and secondary, and crystal growth are far reaching and complex. An unagitated solution can, in general, be cooled further before onset of nucleation, because the overall result of mixing is a decrease in the width of the metastable zone. In a mixed solution at constant supersaturation, with no crystals yet present, mixing intensity can reduce induction time – the time elapsed before crystals first appear. Induction time decreases up to a critical speed and then remains unchanged (Karpinski & Wey, 2001; Mullin, 2001; Tung *et al.*, 2009).

Mixing actively generates secondary nuclei through crystal-crystal, crystal-vessel and crystal-impeller impacts. A greater number of nuclei generated in agitated systems leads to a decrease in the ultimate crystal size of the product. Mixing also has an intensity dependent effect on the mass transfer rate of growing crystals with a growth limiting outcome (Tung *et al.*, 2009).

It has been found that stirring can reproducibly affect the chiral symmetry of crystallization products. If a substance crystallizes as an equal mixture of dextro and levo crystals when unstirred, its chiral symmetry can be disrupted by stirring (Kondepudi *et al.*, 1990; McBride & Carter, 1991). This phenomenon has been attributed to the effect of stirring on secondary nucleation (Kondepuddi & Sabanayagam, 1994).

2.1.5 Pressure

Almost all polymorph screening studies and industrial crystallization processes are performed under ambient pressure, but low pressure is also often applied to increase the rate of solvent evaporation. As with rapid cooling, rapid evaporation leads to higher nucleation rates.

Only recently have researchers turned their attention to high pressure recrystallization from solution as an added dimension in the search for new pharmaceutical polymorphs and solvates. High pressure encourages denser structures in which molecules must pack more efficiently and this means that changes in relative orientations are likely to occur. This gives rise to different themes of molecular interactions, the strengths of which are in turn sensitive to distance and therefore the effects of pressure. The interactions between solute and solvent molecules are also modified by pressure, changing the solubility of a given polymorph or solvate. In some instances, the differences in solubility between two forms might also change, thereby encouraging recrystallization of one form at the expense of another (Fabbiani et al., 2004).

2.1.6 Moisture and humidity

When recrystallizing from hygroscopic organic solvents, care should be taken to use only properly dried solvents (Table 3) or newly opened containers from reputable suppliers. If a crystallization process generates a metastable intermediary form, it can absorb water when exposed to moisture and change into a hydrate. Water molecules, because of their small size and multidirectional hydrogen bonding capabilities, are particularly suited to fill structural voids (Manek & Kolling, 2004). If a crystallization system is open to atmospheric conditions, as is likely to be the case with evaporative crystallization, hygroscopic solvents will absorb moisture from the air if present. Should an evaporative crystallization process be sensitive to the presence of moisture, additional measures will have to be put in place to eliminate atmospheric humidity.

Organic Solvent	Drying Agent
Alcohols	Anhydrous potassium carbonate; anhydrous magnesium or calcium sulphate; calcium oxide.
Alkyl halides Aryl halides	Anhydrous calcium chloride; anhydrous sodium, magnesium or calcium sulphate; phosphorous pentoxide.
Saturated and aromatic hydrocarbons Ethers	Anhydrous calcium chloride; anhydrous calcium sulphate; metallic sodium; phosphorous pentoxide.
Aldehydes	Anhydrous sodium, magnesium or calcium sulphate.
Ketones	Anhydrous sodium, magnesium or calcium sulphate; anhydrous potassium carbonate.
Organic bases (amines)	Solid potassium or sodium hydroxide; calcium oxide; barium oxide.
Organic acids	Anhydrous sodium, magnesium or calcium sulphate.
Halogenated solvents*	Calcium hydride; phosphorous pentoxide.

* Never dry halogenated solvents with alkali metals or alkali metal hydrides as this can cause violent explosions.

Table 3. Common drying agents for organic solvents. (Adapted, with permission of Indian Streams Research Journal, from Shinde et al., 2011.)

2.1.7 Addition of salts, polymers or antisolvents

The addition of salts, polymers or antisolvents can be used to create supersaturation of the solution by decreasing the solute solubility in the crystallization medium. Ions, polymeric molecules, or other substances introduced into the crystallization solvent can also act as impurities for growing crystals. These substances may get adsorbed in the crystal lattice of a growing crystal and disturb the regular and repeating arrangements of the crystal, creating defects and leading to polymorphic modifications. Impurities, and surfactants in particular, can also inhibit crystal growth at certain crystal faces, resulting in crystal habit changes (Tiwary, 2006).

When selecting suitable solvents for antisolvent crystallization, one should select a pair that is miscible. The solute must be more soluble in one solvent and less soluble in the other (antisolvent). A water-soluble API will most likely be more soluble in a polar solvent and less soluble in a non-polar solvent (Table 4). The opposite holds true for a poorly water-soluble API. Antisolvent crystallization is performed by dissolving the API in the solvent and then gradually adding an antisolvent. This results in a decrease of solute solubility and an increase in supersaturation – much like cooling crystallization (Nonoyama *et al.*, 2006; Stieger *et al.*, 2010a; Zhou *et al.*, 2006). Reverse addition is a variation of antisolvent recrystallization whereby the API solution is added to the antisolvent. The resulting rapid increase in supersaturation leads to swift nucleation and the precipitation of very fine particles (Tung *et al.*, 2009). The composition of the crystallization medium can affect both the crystal form and crystal habit of the product (Stieger *et al.*, 2010a).

Solvent 1 (more polar)		Solvent 2 (less polar)	
Solvent	Dielectric constant (ε)	Solvent	Dielectric constant (ε)
Water	78.3	Ethanol	24.3
Water	78.3	Acetone	20.7
Methanol	32.6	Dichloromethane	9.08
Ethanol	24.3	Acetone	20.7
Acetone	20.7	Diethyl ether	4.34
Acetone	20.7	Petroleum ether	1.90
Diethyl ether	4.34	Hexane	1.89
Ethyl acetate	6.02	Cyclohexane	1.97
Ethyl acetate	6.02	Petroleum ether	1.90
Dichloromethane	9.08	Petroleum ether	1.90
Toluene	2.38	Petroleum ether	1.90

Table 4. Common solvent-antisolvent pairs. (Adapted, with permission of the author, from Skonieczny, 2009.)

2.2 Novel strategies and new trends in solvent-based crystallization

One of the biggest challenges for pharmaceutical researchers is finding all the solid-state forms of an API that can exist at ambient conditions. There is no single method for producing all conceivable forms and a particular polymorph may go undetected for many years. Scientists are continually searching for novel ways of uncovering hidden solid-states. A few of these methods are briefly discussed below.

2.2.1 Laser-induced crystallization

Non-photochemical laser-induced nucleation (NPLIN) is a crystallization technique that can affect both nucleation rate and the crystal form produced. Laser pulses act predominantly on pre-existing molecular clusters by assisting in the organization of pre-nucleating clusters and embryos into nuclei (Figure 2), leading to dramatically increased nucleation rates for supersaturated solutions. It is believed that the plane-polarized light aligns the pre-nucleating clusters and thereby reduces the entropic barrier to the free energy of activation for critical nucleus formation (Banga *et al.*, 2004; Garetz *et al.*, 1996; Rodríguez-Spong *et al.*, 2004; Zaccaro *et al.*, 2001). This method has not yet been applied to pharmaceuticals, but it should produce similar results as for other organic substances.

2.2.2 Capillary crystallization

In order to access metastable forms of an API, high levels of supersaturation are often required. Capillary tubes as recrystallization vessels are ideal for manipulating the metastable zone width through slow evaporation and because the small volumes of solution isolate heterogeneous nucleants, and reduce turbulence and convection. This technique offers the additional advantage that the crystals need not be removed from the capillary tube prior to characterization by single crystal- or powder X-ray diffraction (Banga *et al.*, 2004; Rodríguez-Spong *et al.*, 2004).

2.2.3 Sonocrystallization

This technique utilizes ultrasound to increase nucleation rate, but it is also an effective means of crystal size reduction that eliminates many of the disadvantages associated with mechanical size reduction. Sonic waves give rise to a phenomenon called cavitation – the formation of bubbles that decrease in size until a critical size is reached, leading to collapse and the formation of cavities. Cavitation provides energy that accelerates the nucleation process (Banga *et al.*, 2004; Kim *et al*, 2003).

3. Non-solvent based recrystallization

Much emphasis has been placed on the importance of solvent based recrystallization in the production of different solid-state forms of APIs, but non-solvent based recrystallization is equally important to industry and researchers alike. Recrystallization through solid-state transitions affects not only the production of APIs, but also the stability of the final product. Additionally, physical vapor deposition (PVD) recrystallization offers yet more opportunities for the preparation of polymorphs.

3.1 Recrystallization through solid-state transitions

Polymorphs may be obtained through solid-state transition by recrystallization of metastable polymorphs and amorphous forms, or through desolvation of solvates (including hydrates). No solid-state transition will take place unless it is thermodynamically favored and solids will always tend to transform to their lowest energy state – the most stable form. Stability is of major concern where APIs are concerned, but it is also true that stable forms are less soluble than their metastable counterparts and solubility of an API often affects its bioavailability, and therefore also its efficacy (Lipinsky *et al.*, 1997). This necessitates a compromise between stability and solubility: we search for a metastable form with improved solubility that is still relatively stable enough to withstand the rigors of pharmaceutical processing and will yield a product with acceptable shelf-life.

3.1.1 Transformation of polymorphs

Polymorphism, occurring in a single-component crystalline molecular solid, may display monotropism or enantiotropism or both (Figure 8). In a monotropic system (A and C, B and C), only one polymorph is stable below the melting point and a phase change from the metastable form (A and B) to the stable form (C) is irreversible. For enantiotropic systems (A and B), a reversible phase transition is observed at a definite transition temperature ($T_{t,A-B}$), where the free energy curves intersect before the melting point (T_m). At temperatures and pressures below $T_{t,A-B}$ form A will be stable (lower free energy and solubility), whilst in the temperature and pressure range between $T_{t,A-B}$ and $T_{m,B}$ form B is more stable. Polymorphs in an enantiotropic system are referred to as enantiotropes and those in a monotropic system are monotropes (Bernstein 2002; Grant 1999).

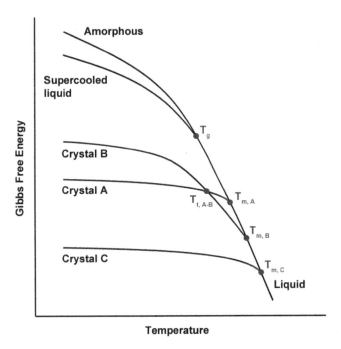

Temperature

Fig. 8. Gibbs free energy curves for a hypothetical single-component system that exhibits crystalline and amorphous phase transitions. Monotropic systems (A and C, B and C), enantiotropic system (A and B) with a transition temperature T_t, and an amorphous and super-cooled liquid with a glass transition temperature T_g. Melting points, T_m, for the crystalline phases are shown at the intersection of curves for the crystalline and liquid states. (Adapted, with permission of the authors, from Shalaev & Zografi, 2002.)

3.1.1.1 Effects of pharmaceutical processing

Unintentional solid-state conversion of polymorphs sometimes occurs upon exposure to the energetics of pharmaceutical processing and a variety of phase conversions are possible when APIs are exposed to milling, wet granulation, oven drying, and compaction (Brittain & Fiese, 1999):

Grinding or milling is often the last step in the production of bulk APIs and it is performed to reduce particle size and improve particle size homogeneity. Milling can impart a significant amount of energy on a solid and could potentially lead to a full or partial polymorphic conversion or generation of an amorphous substance (or at least a degree of amorphous content). Amorphous forms, being metastable, can in turn reconvert to a crystalline state which may differ from the original product.

Wet granulation, used to improve powder flow and blend homogeneity, is a step that often precedes the production of tablets. During this process, the API is exposed to a solvent (water or an organic solvent with low toxicity, like ethanol) and is once more prone to undergoing solvent-mediated transformations. Exposure to humidity can create similar conditions which could lead to a hydrate being formed.

Drying of APIs is typically achieved with heat and moving air. The possible ramifications of temperature change on polymorph stability have been discussed in the previous section. Drying conditions need to be carefully controlled to avoid possible transformations.

Although not generally a common occurrence, compaction (during tablet manufacturing) can potentially cause metastable polymorphs to convert to the stable form. This can be attributed to the energy applied to bring about compaction. The extent of transformation is dependent on the zone of the tablet, the pressure applied, the compression temperature and the particle size of the API powder.

3.1.1.2 High-pressure polymorphic transitions

Solid-state pressure-induced structural changes in molecular crystals can cause polymorphic transitions (Boldyreva, 2003). However, unlike with high-pressure solvent recrystallization (please refer to section 2.1.5) which is experimentally and mechanistically similar, the conversion tends to be only partial and the data are poor. It is thought that even though the application of high pressure to larger organic molecules may thermodynamically favor the adoption of a new polymorphic form, the solid-state has a substantial kinetic barrier to overcome before the molecules are mobile enough to rearrange (Fabbiani et al., 2004).

3.1.2 Desolvation of hydrates and solvates

With the exception of hydrates that are often used in pharmaceutical dosage forms, most solvates are the penultimate solid form in the production of APIs (Byrn et al., 1999). Solvates may be desolvated by removing the recrystallization solvent and exposing the crystals to air at ambient temperature, leading to a decrease in vapor pressure of the solvent. If a solvate is stable under these conditions, it may be dried under vacuum or in an oven at mild temperatures.

Desolvation of a solvate generally results in one of the following forms (Byrn *et al.*, 1999):

- An unsolvated polymorph with a crystal structure different to that of the solvate.
- An unsolvated polymorph with a crystal structure that is the same as that of the solvate.
- An amorphous material that may or may not recrystallize.

Sometimes, a polymorphic transition may mimic the first of these scenarios. Nevirapine's metastable Form IV is isostructural to a series of solvates prepared from primary alcohols (Stieger *et al.*, 2010b) except that its structure contains no solvent molecules. When removed from its recrystallization medium, Form IV will spontaneously and rapidly convert to the stable Form I in exactly the same way as the aforementioned solvates (Stieger *et al.*, 2009).

3.1.3 Crystallization from amorphous phases

Amorphous pharmaceutical solids may be prepared by common pharmaceutical processes including melt quenching, freeze- and spray-drying, milling, wet granulation and desolvation of solvates (Yu, 2001). Glasses are most often produced from a melt of an API. If crystallization does not occur on cooling the melt below the melting point (T_m) of the crystalline phase, a supercooled liquid is obtained. Further cooling induces a change to a glassy state at the glass transition temperature (T_g), accompanied by a dramatic decrease in molecular mobility and heat capacity. In contrast to melting point, the glass transition temperature of a compound can fluctuate with operating conditions (including the rate of cooling/heating) and as a function of the history of the sample. This indicates that the glass transition is a thermal event affected by kinetic factors (Petit & Coquerel, 2006).

An amorphous or glass form is often an intermediate in the production of crystalline APIs and such processes can be useful for overcoming specific kinds of activation energies. Thermodynamically, amorphous solids are out-of-equilibrium states that contain an excess of stored energy, with reference to the crystalline phases (Figure 9), making them unstable by definition. The excess energy can be released either completely through crystallization or partially by means of irreversible relaxation processes (Petit & Coquerel, 2006).

Crystallization from amorphous solids, as for any crystallization, involves successive nucleation and growth steps. The crystallization rate may be affected by numerous, possibly interdependent, parameters notably including temperature and plasticizers (of which water is one). When the temperature of a supercooled liquid decreases from the melting point to the glass transition temperature, the nucleation rate increases exponentially whereas molecular mobility required for growth decreases exponentially. A maximum crystallization rate therefore occurs between preferred temperatures of nucleation and growth. Cooling rate also affects nucleation, with slow cooling allowing the maintenance of a steady-state nucleation rate. Rapid cooling, on the other hand, prevents full development of nucleation and thereby facilitates glass formation (Petit & Coquerel, 2006; Yu, 2001).

Amorphous solids offer certain pharmaceutical advantages over their crystalline counterparts. They are more soluble, have higher dissolution rates, and in some cases they may even have better compression characteristics than corresponding crystals. Therefore, we may not necessarily want amorphous APIs to recrystallize. Some pharmaceuticals have a natural tendency to exist as amorphous solids, while others require deliberate prevention of recrystallization to enter, and remain in, the amorphous state. Contemporary research into the stabilization of amorphous pharmaceuticals focuses on three key areas: (1) the use of additives for the stabilization of labile substances (e.g. proteins and peptides) during processing and storage; (2) prevention of crystallization of excipients that must remain amorphous to perform their intended functions; and (3) determining the appropriate storage conditions for optimal stability of amorphous materials (Yu, 2001).

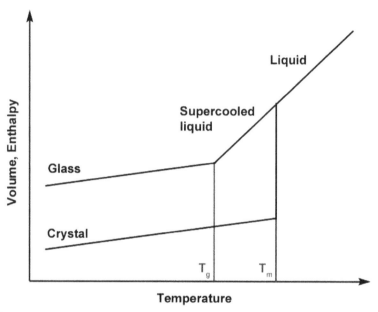

Temperature

Fig. 9. Schematic representation of enthalpy or volume variations as a function of temperature for condensed materials. (Adapted, with permission of John Wiley and Sons, from Petit & Coquerel, 2006.)

3.2 Recrystallization through Physical Vapor Deposition

Physical vapor deposition (PVD) is an atomistic deposition process in which material is vaporized from a solid or liquid source in the form of atoms or molecules and transported in the form of vapor through a low pressure environment to the substrate where it condenses (Mattox, 1998). During the PVD process, molecules unpack from the original crystal lattice and then recrystallize in the new lattice (Byrn *et al.*, 1999).

The research into pharmaceutical applications of PVD has, to date, largely been limited to the production of amorphous phases – particularly ones with enhanced stability. Although we have long known that sublimation may be used for the production of polymorphs, comparatively few papers have been published on recrystallization *via* PVD. This is surprising when one considers that approximately two-thirds of organic compounds sublime (Guillory, 1999) and most commercially available APIs are crystalline.

Studies on the sublimation-based recrystallization of the following APIs and organic compounds have been published: anthranilic acid (Carter & Ward, 1994); 9,10-anthraquinone-2-carboxylic acid (Tsai *et al.*, 1993); 1,1-bis(4-hydroxyphenyl)cyclohexane

(Sarma *et al.*, 2006); caffeine (Griesser *et al.*, 1999; Carlton, 2011); carbamazepine (Griesser *et al.*, 1999; Zeitler *et al.*, 2007); 1,3-dimethyluracil (Sakiyama & Imamura, 1989); ibuprofen (Perlovich *et al.*, 2004;); malonamide (Sakiyama & Imamura, 1989); nevirapine (Figure 10) (Stieger & Liebenberg, 2009); stanozolol (Karpinska *et al.*, 2011); and theophylline (Fokkens *et al.*, 1983; Griesser *et al.*, 1999).

Fig. 10. Micro-spherical aggregates of nano-crystalline nevirapine prepared by PVD (A), the individual crystallites of which could be obtained through light grinding (B).

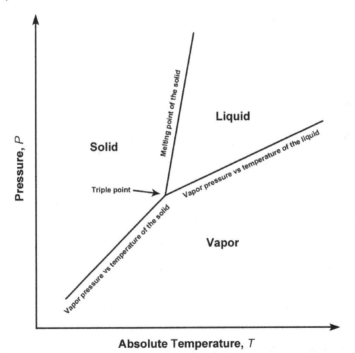

Fig. 11. Pressure-temperature diagram of a one-component crystalline solid for which only one solid phase exists. (Adapted, with permission of Dover Publications, from Ricci, 1966.)

Figure 11 demonstrates how transformations mediated by the vapor phase are highly dependent on vapor pressure and therefore also on temperature (Byrn *et al.*, 1999). Vapor of an API can be generated by its liquid phase once melted (evaporation) or by its solid phase (sublimation). The vast majority of pharmaceutical actives are organic compounds and, as such, they are sensitive to heat degradation. It is well-known that many APIs degrade when melted. It follows that pharmaceutical PVD processes should ideally operate at low pressure and temperature to obtain vapor through sublimation, using compounds with sufficient thermal stability.

Figure 12 illustrates how it is possible to obtain different enantiotropic polymorphs through vapor deposition. At a pressure sufficiently low for sublimation to take place the solid API becomes a vapor, at which point it is no longer Solid A or B because polymorphism is a solid-state phenomenon. The temperature at which the vapor phase recrystallizes determines which polymorph is obtained (provided the pressure is sufficiently low). It can generally be assumed that unstable/metastable polymorphs (in this example, Solid A) will form preferentially at lower temperatures and stable polymorphs (Solid B) can be expected at higher temperatures (Guillory, 1999). This is why the temperature, and distance from the solid, of the collection surface is so important in PVD systems.

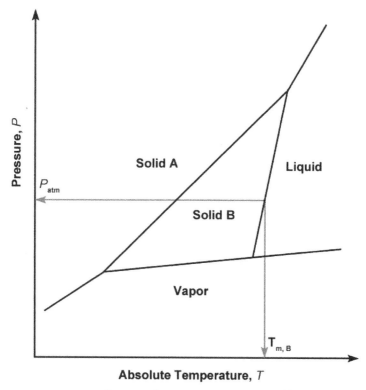

Fig. 12. Pressure-temperature diagram of a one-component crystalline solid with enantiotropic behavior. (Adapted, with permission of John Wiley and Sons, from Lohani & Grant, 2006.)

It has, incidentally, been found that recrystallization from the vapor phase can also be affected or controlled by the nature of the substrate (Carter & Ward, 1994) as has been shown for solvent recrystallization (Rodríguez-Spong et al., 2004).

4. Conclusion

Polymorphs of active pharmaceutical ingredients can be obtained though recrystallization processes based on solvents, solid-state transitions and vapor deposition. When the possible number of experimental variations on each of these is considered, it is easy to become overwhelmed by the task of comprehensive polymorph screening. Indeed, it is no wonder that some polymorphs go undetected for decades or longer. Various semi-automated and high-throughput methods have been developed to assist in this daunting task. The efforts of researchers laboring towards automation are, however "subverted" by colleagues who continually find novel ways of accessing hidden polymorphs! There is, to date, no single method that can identify all possible solid forms of an API.

5. Acknowledgement

We thank North-West University (Potchefstroom Campus) and the National Research Foundation of South Africa for providing research support.

6. References

Banga, S., Chawla, G. & Bansal, A.K. 2004. New Trends in the Crystallisation of Active Pharmaceutical Ingredients. *Business Briefing: Pharmagenerics*, p. 1-5.

Beckman, W. 2000. Seeding the Desired Polymorph: Background, Possibilities, Limitations, and Case Studies. *Organic Process Research & Development*, 4:372-383.

Bernstein, J. 2002. *Polymorphism in Molecular Crystals*. Oxford:Clarendon Press. 410 p.

Blagden, N., De Matas, M., Gavan, P.T. & York, P. 2007. Crystal Engineering of Active Pharmaceutical Ingredients to Improve Solubility and Dissolution Rates. *Advanced Drug Delivery Reviews*, 59(7):617-630.

Boldyreva, E.V. 2003. High-Pressure-Induced Structural Changes in Molecular Crystals Preserving the Space Group Symmetry: Anisotropic Distortion/Isosymmetric Polymorphism. *Crystal Engineering*, 6(4):235-254.

Boistelle, R. & Astier, J.P. 1988. Crystallization Mechanisms in Solution. *Journal of Crystal Growth*, 90(1-3):14-30.

Braatz, R.D. 2002. Advanced Control of Crystallization Processes. *Annual Reviews in Control*, 26(1):87-99.

Brittain, H.G. & Fiese, E.F. 1999. *Effects of Pharmaceutical Processing on Drug Polymorphs and Solvates*. (*In* Brittain, H.G. Ed. Polymorphism in Pharmaceutical Solids. Volume 95: Drugs and the Pharmaceutical Sciences. New York: Marcel Dekker.) p. 331-361.

Brittain, H.G. & Grant, D.J.W. 1999. *Effects of Polymorphism and Solid-State Solvation on Solubility and Dissolution Rate*. (*In* Brittain, H.G. Ed. Polymorphism in Pharmaceutical Solids. Volume 95: Drugs and the Pharmaceutical Sciences. New York: Marcel Dekker.) p. 279-330.

Byrn, S.R., Pfeiffer, R.R. & Stowell, J.G. 1999. *Solid-State Chemistry of Drugs*. 2nd edition. West Lafayette: SSCI. 574 p.

Capes, J.S. & Cameron, R.E. 2007. Effect of Polymer Addition on the Contact Line Crystallization of Paracetamol. *CrystEngComm*, 9:84-90.

Carlton, R.A. 2011. *Pharmaceutical Microscopy*. New York: Springer. 322 p.

Carter, P.W. & Ward, M.D. 1994. Directing Polymorph Selectivity During Nucleation of Anthranilic Acid on Molecular Substrates. *Journal of the American Chemical Society*, 116(2):769-770.

Carstensen, J.T., Ertell, C. & Geoffrey, J. 1993. Physico-Chemical Properties of Particulate Matter. *Drug Development and Industrial Pharmacy*, 19(1-2):195-219.

Doherty, C. & York, P. 1988. Frusemide Crystal Forms: Solid State and Physicochemical Analyses. *International Journal of Pharmaceutics*, 47(1-3):141-155.

Douillet, J., Stevenson, N., Lee, M., Mallet, F., Ward, R., Aspin, P., Dennehy, D.R. & Camus, L. 2011. Development of a Solvate as an Active Pharmaceutical Ingredient: Developability, Crystallisation and Isolation Challenges. *Journal of Crystal Growth*, article in press.

Fabbiani, F.P.A., Allan, D.R., David, W.I.F., Moggach, S.A., Parsons, S. & Pulham, C.R. 2004. High-Pressure Recrystallization – A Route to New Polymorphs and Solvates. *CrystEngComm*, 6(82):504-511.

Florence, A.T. & Attwood, D. 2006. *Physicochemical Principles of Pharmacy*. 4th edition. London: Pharmaceutical Press. 492 p.

Fokkens, J.G., Van Amelsfoort, J.G.M., De Blaey, C.J., De Kruif, C.G. & Wilting, J. 1983. A Thermodynamic Study of the Solubility of Theophylline and its Hydrate. *International Journal of Pharmaceutics*, 14(1):79-93.

Fujiwara, M., Nagy, Z.K., Chew, J.W. & Braatz, R.D. 2005. First-Principles and Direct Design Approaches for the Control of Pharmaceutical Crystallization. *Journal of Process Control*, 15(5):493-504.

Garetz, B.A., Aber, J.E., Goddard, N.L., Young, R.G. & Myerson, A.S. 1996. Nonphotochemical, Polarization-Dependent, Laser-Induced Nucleation in Supersaturated Aqueous Urea Solutions. *Physical Review Letters*, 77:3475-3476.

Grant, D.J.W. 1999. *Theory and Origin of Polymorphism*. (*In* Brittain, H.G. Ed. Polymorphism in Pharmaceutical Solids. Volume 95: Drugs and the Pharmaceutical Sciences. New York: Marcel Dekker.) p. 1-33.

Griesser, U.J. 2006. *The Importance of Solvates*. (*In* Hilfiker, R. Ed. Polymorphism in the Pharmaceutical Industry. Germany: Wiley-VCH.) p. 211-233.

Griesser, U.J., Szelagiewicz, M., Hofmeier, U.C., Pitt, C. & Cianferani, S. 1999. Vapor Pressure and Heat of Sublimation of Crystal Polymorphs. *Journal of Thermal Analysis and Calorimetry*, 57(1):45-60.

Guillory, J.K. 1999. *Generation of Polymorphs, Hydrates, Solvates and Amorphous Solids*. (*In* Brittain, H.G. Ed. Polymorphism in Pharmaceutical Solids. Volume 95: Drugs and the Pharmaceutical Sciences. New York: Marcel Dekker.) p. 183-225.

Haleblian, J.K. 1975. Characterization of Habits and Crystalline Modifications of Solids and their Pharmaceutical Applications. *Journal of Pharmaceutical Sciences*, 64:1269-1288.

Karpinska, J., Erxleben, A. & McArdle, P. 2011. 17β-Hydroxy-17α-methylandrostanol[3,2-c]pyrazole, Stanozolol: The Crystal Structures of Polymorphs 1 and 2 and 10 Solvates. Crystal Growth & Design, 11(7):2829-2838.

Kim, S., Wei, C. & Kiang, S. 2003. Crystallization Process Development of an Active Pharmaceutical Ingredient and Particle Engineering *via* the Use of Ultrasonics and Temperature Cycling. *Organic Process Research Development*, 7(6):997-1001.

Karpinski, P.H. & Wey, J.S. 2001. *Precipitation Processes.* (*In* Myerson, A.S. *Ed.* Handbook of Industrial Crystallization. 2nd edition. Woburn: Butterworth-Heinemann.) p. 141-159.

Kondepuddi, D.K., Kaufmann, R.J. & Singh, N. 1990. Chiral Symmetry Breaking in Sodium Chlorate Crystallization. *Science*, 250(4983):975-976.

Kondepuddi, D.K. & Sabanayagam, C. 1994. Secondary Nucleation that Leads to Chiral Symmetry Breaking in Stirred Crystallization. *Chemical Physics Letters*, 217(4):364-368.

Kuzmenko, I., Rapaport, H., Kjaer, K., Als-Nielsen, J., Weissbuch, I., Lahav, M. & Leiserowitz, L. 2001. Design and Characterization of Crystalline Thin Film Architectures at the Air-Liquid Interface: Simplicity to Complexity. *Chemical Reviews*, 101(6):1659-1696.

Lipinski, C.A., Lombardo, F., Dominy, B.W. & Feeney, P.J. 1997. Experimental and Computational Approaches to Estimate Solubility and Permeability in Drug Discovery and Development Settings. *Advanced Drug Delivery Reviews*, 46:3-26.

Lohani, S. & Grant, J.W.G. 2006. *Thermodynamics of Polymorphs.* (*In* Hilfiker, R. *Ed.* Polymorphism in the Pharmaceutical Industry. Germany: Wiley-VCH.) p. 21-42.

Manek, R.V. & Kolling, W.M. 2004. Influence of Moisture on the Crystal Forms of Niclosamide Obtained from Acetone and Ethyl Acetate. *AAPS PharmSciTech*, 5(1):101-108.

Mattox, D.M. 2010. *Handbook of Physical Vapor Deposition (PVD) Processing.* 2nd edition. Oxford: William Andrew. 792 p.

McBride, J.M. & Carter, R.L. 1991. Spontaneous Resolution by Stirred Crystallization. *Angewandte Chemie*, 30(3):293-295.

Mullin, J.W. 2001. *Crystallization.* 4th edition. Oxford: Butterworth-Heinemann. p. 322.

Ostwald, F.W. 1897. Studien uber die Bildung und Umwandlung fester Korper. *Zeitschrift für Physikalische Chemie*, 22:289.

Nonoyama, N., Hanaki, K. & Yabuki, Y. 2006. Constant Supersaturation Control of Antisolvent-Addition Batch Crystallization. *Organic Process Research & Development*, 10(4):727-732.

Perlovich, G.L., Kurkov, S.V., Hansen, L.K. & Bauer-Brandl, A. 2004. Thermodynamics of Sublimation, Crystal Lattice Energies, and Crystal Structures of Racemates and Enantiomers: (+)- and (-)-Ibuprofen. *Journal of Pharmaceutical Sciences*, 93(3):654-666.

Petit, S. & Coquerel, G. 2006. *The Amorphous State.* (*In* Hilfiker, R. *Ed.* Polymorphism in the Pharmaceutical Industry. Germany: Wiley-VCH.) p. 259-285.

Ricci, J.E. 1966. *The Phase Rule and Heterogeneous Equilibrium.* New York: Dover Publications. 505 p.

Rodríguez-Homedo, N., Kelly, R.C., Sinclair, B.D. & Miller, J.M. 2006. *Crystallization: General Principles and Significance on Product Development.* (*In* Swarbrick, J. *Ed.* Encyclopedia of Pharmaceutical Technology. 3rd edition. New York: Informa Healthcare.) p. 834-857.

Rodríguez-Spong, B., Price, C.P., Jayasankar, A., Matzger, A.J. & Rodríguez-Homedo, N. 2004. General Principles of Pharmaceutical Solid Polymorphism: A Supramolecular Perspective. *Advanced Drug Delivery Reviews*, 56:241-274.

Rouhi, A.M. 2003. The Right Stuff. *Chemical & Engineering News*, 81(8):32-35.

Sakiyama, M. & Imamura, A. 1989. Thermoanalytical Characterization of 1,3-Dimethyluracil and Malonamide Crystals. *Thermochimica Acta*, 142(2):365-370.

Sarma, B., Roy, S. & Nangia, A. 2006. Polymorphs of 1,1-Bis(4-hydroxyphenyl)cyclohexane and Multiple Z' Crystal Structures by Melt and Sublimation Crystallization. *Chemical Communications*, 47:4918-4920.

Shalaev, E. & Zografi, G. 2002. *The Concept of 'Structure' in Amorphous Solids from the Perspective of the Pharmaceutical Sciences.* (*In* Levine, H. *Ed.* Amorphous Food and Pharmaceutical Systems. Cambridge: The Royal Society of Chemistry.) p. 11-30.

Shekunov, B.Y. & York, P. 2000. Crystallization Processes in Pharmaceutical Technology and Drug Delivery Design. *Journal of Crystal Growth*, 211(1-4):122-136.

Shinde, K.R., Patil, S.D. & Dhake, A.S. 2011. Recrystallization. *Indian Streams Research Journal*, 1(4).

Skonieczny, S. 2009. Crystallization. *University of Toronto course notes for CHM249.* p. 1-18.

Stieger, N., Caira, M.R., Liebenberg, W., Tiedt, L.R., Wessels, J.C. & De Villiers, M.M. 2010a. Influence of the Composition of Water/Ethanol Mixtures on the Solubility and Recrystallization of Nevirapine. *Crystal Growth & Design*, 10(9):3859-3868.

Stieger, N. & Liebenberg, W. 2009. Method for Increasing the Solubility of a Transcriptase Inhibitor Composition. *Patent application PCT/IB2010/055077* (priority date 2009-11-10).

Stieger, N., Liebenberg, W. & Caira, M.R. 2009. Method of Producing a Polymorph Form. *Patent application PCT/IB2010/055808* (priority date 2009-12-17).

Stieger, N., Liebenberg, W., Wessels, J.C., Samsodien, H. & Caira, M.R. 2010b. Channel Inclusion of Primary Alcohols in Isostructural Solvates of the Antiretroviral Nevirapine: an X-Ray an Thermal Analysis Study. *Structural Chemistry*, 21(4):771-777.

Stoica, C., Verwer, P., Meekes, H., Van Hoof, P.J.C.M., Kaspersen, F.M. & Vlieg, E. 2004. Understanding the Effect of a Solvent on the Crystal Habit. *Crystal Growth & Design*, 4(4):765-768.

Threlfall, T. 2000. Crystallisation of Polymorphs: Thermodynamic Insight into the Role of Solvent. *Organic Process Research & Development*, 4(5):384-390.

Tiwary, A.K. 2006. *Crystal Habit Changes and Dosage Form Performance.* (*In* Swarbrick, J. *Ed.* Encyclopedia of Pharmaceutical Technology. 3rd edition. New York: Informa Healthcare.) p. 822.

Togkalidou, T., Tung, H., Sun, Y., Andrews, A. & Braatz, R.D. 2002. Solution Concentration Prediction for Pharmaceutical Crystallization Processes Using Robust Chemometrics and ATR FTIR Spectroscopy. *Organic Process Research Development*, 6(3):317-322.

Tros de Ilarduya, M.C., Martin, C., Goñi, M.M. & Martinez-Uhárriz, M.C. 1997. Dissolution Rates of Polymorphs and Two New Pseudopolymorphs of Sulindac. *Drug Development and Industrial Pharmacy*, 23:1095-1098.

Tsai, S., Kuo, S. & Lin, S. 1993. Physicochemical Characterization of 9,10-anthraquinone-2-carboxilic acid. *Journal of Pharmaceutical Sciences*, 82(12):1250-1254.

Tung, H., Paul, E.L., Midler, M. & McCauly, J.A. 2009. *Crystallization of Organic Compounds: An Industrial Perspective.* New Jersey: John Wiley & Sons. 289 p.

Yu, L. 2001. Amorphous Pharmaceutical Solids: Preparation, Characterization and Stabilization. *Advanced Drug Delivery Reviews*, 48(1):27-42.

Zaccaro, J., Matic, J., Myerson, A.S. & Garetz, B.A. 2001. Nonphotochemical, Laser-Induced Nucleation of Supersaturated Aqueous Glycine Produces Unexpected γ-Polymorph. *Crystal Growth & Design*, 1(1):5-8.

Zeitler, J.A., Taday, P.F., Gordon, K.C., Pepper, M. & Rades, T. 2007. New Insights into the Solid-State Transition Mechanism in Carbamazepine Polymorphs by Time-Resolved Terahertz Spectroscopy. *ChemPhysChem*, 8(13):1924-1927.

Zhou, G.X., Fujiwara, M., Woo, X.Y., Rusli, E., Tung, H., Starbuck, C., Davidson, O., Ge, Z. & Braatz, R.D. 2006. Direct Design of Pharmaceutical Antisolvent Crystallization through Concentration Control. *Crystal Growth & Design*, 6(4):892-898.

5

A Mathematical Model for Single Crystal Cylindrical Tube Growth by the Edge-Defined Film-Fed Growth (EFG) Technique

Loredana Tanasie and Stefan Balint
West University of Timisoara
Romania

1. Introduction

1.1 Crystal growth from the melt by E.F.G. technique

Modern engineering does not only need crystals of arbitrary shapes but also plate, rod and tube-shaped crystals, i.e., crystals of shapes that allow their use as final products without additional machining. Therefore, the growth of crystals of specified sizes and shapes with controlled defect and impurity structures are required. In the case of crystals grown from the melt, this problem appears to be solved by profiled-container crystallization as in the case of casting. However, this solution is not always possible, for example growing very thin plate-shaped crystals from the melt (to say nothing of more complicated shapes), excludes container application completely.[Tatarchenko, 1993]

The techniques which allow the shaping of the lateral crystal surface without contact with the container walls are appropriate for the above purpose. In the case of these techniques the shapes and the dimensions of the grown crystals are controlled by the interface and meniscus-shaping capillary force and by the heat- and mass-exchange conditions in the crystal-melt system. The edge-defined film-fed growth (EFG) technique is of this type. Whenever the E.F.G. technique is employed, a shaping device is used (Fig. 1). In the device a capillary channel is manufactured (Fig. 1) in which the melt raises and feeds the growth process. Frequently, a wettable solid body is used to raise the melt column above the shaper, where a thin film is formed. When a wettable body is in contact with the melt, an equilibrium liquid column embracing the surface of the body is formed. The column formation is caused by the capillary forces being present. Such liquid configuration is usually called a meniscus (Fig. 1) and in the E.F.G. technique, its lower boundary (Fig. 1 – point C) is attached to the sharp edge of the shaper.

Let be the temperature of the meniscus upper horizontal section (Fig. 1 – $\overline{AB}$) the temperature of the liquid crystallization. So, above the plane of this section, the melt transforms in solid phase. Now set the liquid phase into upward motion with the constant rate, v, keeping the position of the phase-transition plane invariable by selection of the heat conditions. When the motion starts, the crystallized position of the meniscus will continuously form a solid upward or downward tapering body. In the particular case when the line tangent at the triple point B to the liquid meniscus surface makes a specific angle (angle of growth) with the vertical, the lateral wall of the crystal will be vertical. Thus, the initial body, called the seed, serves to form a meniscus which later on determines the form of the crystallized product, the phase transition position being fixed.

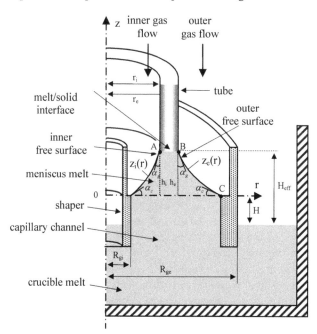

Fig. 1. Prototype tubular crystal growth by E.F.G. method

Based on this description, a conclusion can be drawn that the dimensions and shapes of the specimens being pulled by the E.F.G. technique depend upon the following factors: (i) the shaper geometry; (ii) the pressure of feeding the melt to the shaper; (iii) the crystallization front position; (iv) the seed's shape. The seed's shape is only important for stationary pulling; in this case its cross-section should coincide with the desired product's cross-section. Frequently, especially when complicated profiles are grown, the pulling process is carried out under unstationary conditions by lowering the crystallization surface, which then enhances the dependence of the shapes on the crystal cross-section. With such an approach applied to the pulling process, the dimensions and the shape of the grown crystal are determined by the above-mentioned factors and by the pulling rate-to-crystallization front displacement ratio.[Tatarchenko, 1993].

1.2 Background history of tube growth from the melt by E.F.G. method

The technology of growing tubes can have a significant impact for example on the solar cell technology. The growth of silicon tubes by E.F.G. process was first reported by Erris et al. [Erris et al.,1980]. Tubes were grown with a diameter of 95×10^{-4} [m], wall thickness in the range of $5 \times 10^{-5} - 1 \times 10^{-3}$ [m] at rates up to 2×10^{-3} [m/s]. In [Erris et al.,1980] a theory of tube growth by the E.F.G. process is developed to show the dependence of the tube wall thickness on the growth variables. The theory concerns the calculations of the shape of the liquid-vapor interface (or meniscus) and of the heat flow in the system. The inner and outer meniscus shapes, (Fig.1), are both calculated from Laplace's capillary equation, in which the pressure difference Δp across a point on meniscus is considered to be $\Delta p = \rho \circ g \circ H_{eff} =$ constant, where H_{eff} represents the effective height of the growth interface above the horizontal liquid level in the crucible (Fig.1). According to [Surek et al.,1977], [Swartz et al., 1975], it includes the effects of the viscous flow of the melt in the shaper capillary and in the meniscus film, as well as that of the hydrostatic head. The above approximation for Δp is valid for silicon ribbon growth [Surek et al.,1977], [Kalejs et al., 1990], when $H_{eff} >> h$, where h is the height of the growth interface above the shaper top (i.e. the meniscus height). Another approximation used in [Erris et al.,1980], concerning the meniscus, is that the inner and outer meniscus shapes are approximated by circular segments. With these relatively tight tolerances concerning the menisci in conjunction with the heat flow calculation in the system, the predictive model developed in [Erris et al.,1980] has been shown to be a useful tool in understanding the feasible limits of wall thickness control. A more precise predictive model would require an increase of the acceptable tolerance range introduced by approximation.

Later, this process was scaled up by Kaljes et al. [Kalejs et al., 1990] to grow 15×10^{-2} [m] diameter silicon tubes, and the stress behavior in the grown tube was investigated. It has been realized that numerical investigations are necessary for the improvement of the technology. Since the growth system consists of a small die tip (1×10^{-3} m width) and a thin tube (order of 200×10^{-6} [m] wall thickness) the width of the melt/solid interface and meniscus are accordingly very small. Therefore, it is essential to obtain an accurate solution for the temperature and interface position in this tiny region.

In [Rajendran et al., 1993] an axisymmetric finite element model of magnetic and thermal field was presented for an inductively heated furnace. Later the same model was used to determine the critical parameters controlling silicon carbide precipitation on the die wall [Rajendran et al., 1994]. Rajendran et al. also developed a three dimensional magnetic induction model for an octagonal E.F.G, system. Recently, in [Roy et al., 2000a], [Roy et al., 2000b], a generic numerical model for an inductively heated large diameter Si tube growth system was reported. In [Sun et al., 2004] a numerical model based on multi-block method and multi-grid technique is developed for induction heating and thermal transport in an E.F.G. system. The model is applied to investigate the growth of large octagon silicon tubes of up to 50×10^{-2} m diameter. A 3D dynamic stress model for the growth of hollow silicon polygons is reported in [Behnken et al., 2005]. In [Mackintosh et al., 2006] the challenges fixed in bringing E.F.G. technology into large-scale manufacturing, and ongoing development of furnace designs for growth of tubes for larger wafer production using hexagons with 1×10^{-2} m face widths, and wall thicknesses in he range $250 \times 10^{-6} - 300 \times 10^{-6}$ m is described. In [Kasjanow et al., 2010] the authors present a 3D coupled electromagnetic and thermal modeling of E.F.G. silicon tube growth, successfully validated by experimental tests with industrial installations.

The state of the art at 1993-1994 concerning the calculation of the meniscus shape in general in the case of the growth by E.F.G. method is summarized in [Tatarchenko, 1993]. According to [Tatarchenko, 1993], for the general equation describing the surface of a liquid meniscus possessing axial symmetry, there is no complete analysis and solution. For the general equation only numerical integration was carried out for a number of process parameter values that are of practical interest at the moment. The authors of papers [Borodin&Borodin&Sidorov&Petkov, 1999],[Borodin&Borodin&Zhdanov, 1999] consider automated crystal growth processes based on weight sensors and computers. They give an expression for the weight of the meniscus, contacted with a crystal and shaper of arbitrary shape, in which there are two terms related to the hydrodynamic factor. In [Rosolenko et al., 2001] it is shown that the hydrodynamic factor is too small to be considered in the automated crystal growth and it is not clear what equation (of non Laplace type) was considered for the meniscus surface. Finally, in [Yang et al., 2006] the authors present theoretical and numerical study of meniscus dynamics under symmetric and asymmetric configurations. A meniscus dynamics model is developed to consider meniscus shape and its dynamics, heat and mass transfer around the die-top and meniscus. Analysis reveals the correlations between tube thickness, effective melt height, pull-rate, die-top temperature and crystal environmental temperature.

The purpose of this chapter is the mathematical description of the growth process of a single crystal cylindrical tube grown by the edge-defined film-fed growth (EFG) technique. The mathematical model defined by a set of three differential equations governing the evolution of the outer radius and the inner radius of the tube and of the crystallization front level is the one considered in [Tatarchenko, 1993]. This system contains two functions which represent the angle made by the tangent line to the outer (inner) meniscus surface at the three-phase point with the horizontal. The meniscus surface is described mathematically by the solution of the axi-symmetric Young-Laplace differential equation. The analysis of the dependence of solutions of the Young-Laplace differential equation on the pressure difference across the free surface, reveals necessary or sufficient conditions for the existence of solutions which represent convex or concave outer or inner free surfaces of a meniscus. These conditions are expressed in terms of inequalities which are used for the choice of the pressure difference, in order to obtain a single-crystal cylindrical tube with specified sizes.

A numerical procedure for determining the functions appearing in the system of differential equations governing the evolution is presented.

Finally, a procedure is presented for setting the pulling rate, capillary and thermal conditions to grow a cylindrical tube with prior established inner and outer radius. The right hand terms of the system of differential equations serve as tools for setting the above parameters. At the end a numerical simulation of the growth process is presented.

The results presented in this chapter were obtained by the authors and have never been included in a book concerning this topic.

Since the calculus and simulation in this model can be made by a P.C., the information obtained in this way is less expressive than an experiment and can be useful for experiment planing.

2. The system of differential equations which governs the evolution of the tube's inner radius r_i, outer radius r_e and the level of the crystallization front h

According to [Tatarchenko, 1993] the system of differential equations which governs the evolution of the tube's inner radius r_i, the outer radius r_e and the level of the crystallization front h is:

$$
\begin{cases}
\dfrac{dr_e}{dt} = -v \cdot \tan\left[\overline{\alpha}_e(r_e,h,p_e) - \left(\dfrac{\pi}{2} - \alpha_g\right)\right] \\[2ex]
\dfrac{dr_i}{dt} = v \cdot \tan\left[\overline{\alpha}_i(r_i,h,p_i) - \left(\dfrac{\pi}{2} - \alpha_g\right)\right] \\[2ex]
\dfrac{dh}{dt} = v - \dfrac{1}{\Lambda \cdot \rho_1} \cdot \left[\lambda_1 \cdot G_1(r_e,r_i,h) - \lambda_2 \cdot G_2(r_e,r_i,h)\right]
\end{cases}
\tag{1}
$$

In equations $(1)_1$ and $(1)_2$: v is the pulling rate, $\overline{\alpha}_e(r_e,h,p_e)$ ($\overline{\alpha}_i(r_i,h,p_i)$) is the angle between the tangent line to the outer (inner) meniscus at the three phase point of coordinates (r_e,h) ((r_i,h)) and the horizontal Or axis (Fig.1 b), α_g is the growth angle (Fig. 1), p_e (p_i) is the controllable part of the pressure difference across the free surface given by:

$$
p_e = p_m - p_g^e - \rho_1 \cdot g \cdot H_e \quad (p_i = p_m - p_g^i - \rho_1 \cdot g \cdot H_i)
\tag{2}
$$

where p_m is the hydrodynamic pressure in the melt under the free surface, which can be neglected in general, with respect to the hydrostatic pressure $\rho_1 \cdot g \cdot H_e$ ($\rho_1 \cdot g \cdot H_i$); p_g^e (p_g^i) is the pressure of the gas flow, introduced in order to release the heat from the outer (inner) wall of the tube; H_e (H_i) is the melt column height between the horizontal crucible melt level and the shaper outer (inner) top level (Fig. 1a); ρ_1 is the melt density; g is the gravity acceleration.

The angle $\overline{\alpha}_e(r_e,h,p_e)$ ($\overline{\alpha}_i(r_i,h,p_i)$) fluctuates due to the fluctuations of: the outer (inner) radius r_e (r_i), the level h of the crystallization front and the outer (inner) pressure p_e (p_i)

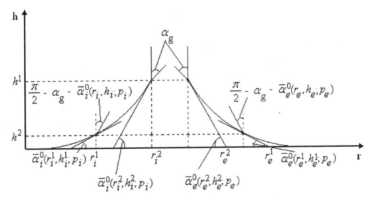

Fig. 2. Fluctuations at the triple point

In the equation $(1)_3$: Λ is the latent melting heat; λ_1, λ_2 are the thermal conductivity coefficients in the melt and the crystal respectively; G_1^j, G_2^j are the temperature gradients at the interface in the melt $(i=1)$ and in the crystal $(i=2)$ respectively, given by the formulas:

$$G_1^j(r_e,r_i,h) = \frac{1}{SINH(\beta_{1j}\cdot h)}\left[\left(T_0 - T_{en}(0) - \frac{v\cdot k}{(F_j)^2\cdot Bi\cdot\chi_1}\right)\cdot\left(-\beta_{1j}\cdot e^{\delta_1\cdot h}\right)+\right.$$
$$\left.\left(T_m - T_{en}(0) + k\cdot h - \frac{v\cdot k}{(F_j)^2\cdot Bi\cdot\chi_1}\right)\cdot\left(\delta_1\cdot SINH(\beta_{1j}\cdot h) + \beta_{1j}\cdot COSH(\beta_{1j}\cdot h)\right)\right] - k$$

(3)

$$G_2^j(r_e,r_i,h) = \frac{1}{SINH(\beta_{2j}\cdot L)}\left[\left(T_m - T_{en}(0) + k\cdot h - \frac{v\cdot k}{(F_j)^2\cdot Bi\cdot\chi_2}\right)\cdot\right.$$
$$\left.\left(\delta_2\cdot SINH(\beta_{2j}\cdot L) - \beta_{2j}\cdot COSH(\beta_{2j}\cdot L)\right) - \frac{v\cdot k}{(F_j)^2\cdot Bi\cdot\chi_2}\cdot\beta_{2j}\cdot e^{-\delta_2\cdot L}\right] - k$$

(4)

where χ_i - the thermal diffusivity coefficient equal to $\dfrac{\lambda_i}{\rho_i\cdot c_i}$, ρ_i - the density, c_i - the heat capacity, Bi - the Biot number equal to $\dfrac{\mu_i\cdot r_e}{\lambda_i}$ ($i=1$ - the melt, $i=2$ - the crystal), $\mu_i = \mu_i^k + \mu_i^r$ - the coefficient of the heat-exchange with environment (μ_i^k - the convective heat-exchange coefficient and μ_i^r - the linearized radiation heat-exchange coefficient), F_j ($j=1,2,3$) the crystal (meniscus) cross – section perimeter – to – its area ratio: $j=1$ and $F_1 = \dfrac{2}{r_e}$, for a thick-walled tube with small inner radius, for which heat is removed from the external surface only, $j=2$ and $F_2 = \dfrac{2\cdot r_e}{r_e^2 - r_i^2}$ for a tube of not to large inner radius for which heat is removed from the external surface only, $j=3$ and $F_3 = \dfrac{2}{r_e - r_i}$ for a tube for which heat is removed from both the outer and inner surfaces([Tatarchenko, 1993], pp. 39-40, 146). T_0 - the melt temperature at the meniscus basis, T_m - melting temperature, $T_{en}(0)$ - the environment temperature at $z=0$, k - the vertical temperature gradient in the furnace, r_e - the outer radius of the tube equal to the upper radius of the outer meniscus, r_i - the inner radius of the tube equal to the upper radius of the inner meniscus, L - the tube length and

$\delta_i = \dfrac{v}{2\chi_i}$, $\beta_{ij} = \sqrt{\dfrac{v^2}{4\chi_i^2} + (F_j)^2\cdot Bi}$, $i=1,2, j=1,2,3$, $SINH$ and $COSH$ are the hyperbolic sine and hyperbolic cosine functions.

Due to the supercooling in this gradients it is assumed that: $T_0 < T_m$, $k>0$, $T_{en}(0) < T_m$.

In the following sections we will show in which way $\overline{\alpha}_e(r_e,h,p_e)$ and $\overline{\alpha}_i(r_i,h,p_i)$ can be found starting from the Young-Laplace equation of a capillary surface in equilibrium.

3. The choice of the pressure of the gas flow and the melt level in silicon tube growth

In a single crystal tube growth by edge-defined film-fed growth (E.F.G.) technique, in hydrostatic approximation, the free surface of a static meniscus is described by the Young-Laplace capillary equation [Finn, 1986]:

$$\gamma \cdot \left(\frac{1}{R_1} + \frac{1}{R_2} \right) = \rho \cdot g \cdot z - p \tag{5}$$

Here γ is the melt surface tension, ρ denotes the melt density, g is the gravity acceleration, $1/R_1, 1/R_2$ denote the mean normal curvatures of the free surface at a point M of the free surface, z is the coordinate of M with respect to the Oz axis, directed vertically upwards, p is the pressure difference across the free surface. To calculate the outer and inner free surface shape of the static meniscus it is convenient to employ the Young-Laplace eq.(5) in its differential form. This form of the eq.(5) can be obtained as a necessary condition for the minimum of the free energy of the melt column [Finn, 1986].For a tube of outer radius $r_e \in \left(\dfrac{R_{gi} + R_{ge}}{2}, R_{ge} \right)$ and inner radius $r_i \in \left(R_{gi}, \dfrac{R_{gi} + R_{ge}}{2} \right)$, the axi-symmetric differential equation of the outer free surface is given by:

$$z'' = \frac{\rho \cdot g \cdot z - p_e}{\gamma} \left[1 + (z')^2 \right]^{3/2} - \frac{1}{r} \cdot \left[1 + (z')^2 \right] \cdot z' \quad \text{for } r \in [r_e, R_{ge}] \tag{6}$$

which is the Euler equation for the energy functional

$$I_e(z) = \int_{r_e}^{R_{ge}} \left\{ \gamma \cdot \left[1 + (z')^2 \right]^{1/2} + \frac{1}{2} \cdot \rho \cdot g \cdot z^2 - p_e \cdot z \right\} \cdot r \cdot dr, \quad z(r_e) = z_e(r_e),$$
$$z(R_{ge}) = z_e(R_{ge}) = 0 \tag{7}$$

The axi-symmetric differential equation of the inner free surface is given by:

$$z'' = \frac{\rho \cdot g \cdot z - p_i}{\gamma} \left[1 + (z')^2 \right]^{3/2} - \frac{1}{r} \cdot \left[1 + (z')^2 \right] \cdot z' \quad \text{for } r \in \left[R_{gi}, r_i \right] \tag{8}$$

which is the Euler equation for the energy functional:

$$I_i(z) = \int_{R_{gi}}^{r_i} \left\{ \gamma \cdot \left[1 + (z')^2 \right]^{1/2} + \frac{1}{2} \cdot \rho \cdot g \cdot z^2 - p \cdot z \right\} \cdot r \cdot dr, \quad z(R_{gi}) = z_i(R_{gi}) = 0, \quad z(r_i) = z_i(r_i) \tag{9}$$

In papers [Balint & Balint, 2009b], [Balint&Balint&Tanasie, 2008], [Balint & Tanasie, 2008] , Balint, Tanasie, 2011] some mathematical theorems and corollaries have been rigorously proven regarding the existence of an appropriate meniscus. These results are presented in Appendixes. In the following we will shown in which way the inequalities can be used for creation of the appropriate meniscus.

3.1 Convex free surface creation

In this section, it will be shown in which way the inequalities presented in Appendix 1 can be used for the creation of an appropriate static convex meniscus by the choice of p_e and p_i [Balint, Tanasie, 2011].

Inequalities (A.1.1) establish the range where the pressure difference p_e has to be chosen in order to obtain a static meniscus with convex outer free surface, appropriate for the growth of a tube of outer radius equal to $\dfrac{R_{ge}}{n}$.

If the pressure difference satisfies (A.1.2), then a static meniscus with convex outer free surface is obtained which is appropriate for the growth of a tube of outer radius $r_e \in \left[\dfrac{R_{ge}}{n}, R_{ge} \right]$.

If the pressure difference satisfies inequality (A.1.4) and the value of p_e is close to the value of the right hand member of the inequality (A.1.4) then a static meniscus with convex outer free surface is obtained which is appropriate for the growth of a tube of outer radius equal to $\dfrac{R_{ge} + R_{gi}}{2}$.

If the pressure difference satisfies inequality (A.1.5), then a static meniscus with convex outer free surface is obtained which is appropriate for the growth of a tube of outer radius in the range $\left[\dfrac{R_{ge}}{n}, \dfrac{R_{ge}}{n'} \right]$.

Theorem 5 (Appendix 1) shows that a static meniscus having a convex outer free surface, appropriate for the growth of a tube of outer radius r_e situated in the range $\left[\dfrac{R_{ge}}{n}, \dfrac{R_{ge}}{n'} \right]$, is stable.

Inequalities (A.1.6) establish the range where the pressure difference p_i has to be chosen in order to obtain a static meniscus with convex inner free surface appropriate for the growth of a tube of inner radius equal to $m \cdot R_{gi}$.

If the pressure difference p_i satisfies (A.1.7) then a static meniscus with convex inner free surface is obtained which is appropriate for the growth of a tube of inner radius $r_i < m \cdot R_{gi}$.

If the pressure difference p_i satisfies the inequality (A.1.9) and the value of p_i is close to the value of the right hand term of the inequality (A.1.9) then a static meniscus with convex inner free surface is obtained which is appropriate for the growth of a tube of inner radius equal to $\dfrac{R_{ge} + R_{gi}}{2}$.

If the pressure difference p_i satisfies inequality (A.1.10) then a static meniscus with convex inner free surface is obtained which is appropriate for the growth of a tube of inner radius which is in the range $\left[m' \cdot R_{gi}, m \cdot R_{gi} \right]$.

Theorem 10 (Appendix 1) shows that a static meniscus having a covex inner free surface appropriate for the growth of a tube of inner radius r_i situated in the range $\left[R_{gi}, \dfrac{R_{gi} + R_{ge}}{2} \right]$ is stable.

For numerical illustrations, the inner radius of the shaper was taken $R_{gi} = 4.2 \times 10^{-3}$ [m] and outer radius of the shaper was chosen $R_{ge} = 4.8 \times 10^{-3}$ [m] [Eriss, 1980]. Computations were performed in **MathCAD 14.** and for **Si** the following numerical values were considered: $\rho = 2.5 \times 10^{3}$ [kg/m³]; $\gamma = 7.2 \times 10^{-1}$ [N/m]; $\alpha_c = 30°$; $\alpha_g = 11°$; g=9.81[m/s²].

To create a convex meniscus appropriate for the growth of a tube having the outer radius r_e^1 equal to $r_e^1 = 4.65 \times 10^{-3}$ [m] ($n_1 = 1.03226$), according to the **Theorem 1** (Appendix 1), p_e has to be chosen in the range: $[-3480.07, -612.35]$ [Pa]. According to the **Corollary 3** (Appendix 1), from this range for the values of p_e smaller than -1702.52 [Pa] the point r_e where $z_e'(r_e) = -\tan\left(\frac{\pi}{2} - \alpha_g\right)$ is close to 4.5×10^{-3} [m]. Hence, we have to find for $p_e = -1702.52$ [Pa] the point r_e for which the above condition is satisfied. This can be made by integrating numerically the following system for $z(R_{ge}) = 0$, $\alpha(R_{ge}) = \alpha_c$ and $p_e = -1702.52$ [Pa] (see Fig. 3):

$$\begin{cases} \dfrac{dz_e}{dr} = -\tan(\alpha_e) \\ \dfrac{d\alpha_e}{dr} = \dfrac{p_e - \rho \cdot g \cdot z_e}{\gamma} \cdot \dfrac{1}{\cos \alpha_e} - \dfrac{1}{r} \cdot \tan \alpha_e \end{cases} \tag{10}$$

Since the obtained r_e is $r_e = 4.609 \times 10^{-3}$ [m], and it is smaller than the desired value $r_e^1 = 4.65 \times 10^{-3}$ [m], the value of p_e has to be chosen in the range $[-3480.07, -1702.52]$ [Pa].

The results of the integrations of the system (10) for $z(R_{ge}) = 0$, $\alpha(R_{ge}) = \alpha_c$ and different values of p_e in this range, are presented in Fig. 4. This figure shows that the outer radius $r_e^1 = 4.65 \times 10^{-3}$ [m] is obtained for $p_e^1 = -2198$ [Pa].

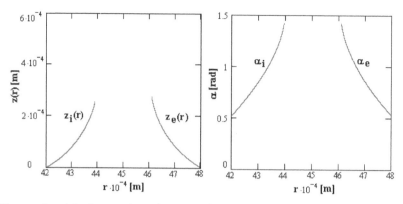

Fig. 3. The results of the integration of systems (10) and (11) for $p_e = -1702.52$ [Pa] and $p_i = -1945.80$ [Pa]

Taking $p_m \approx 0$ [Eriss et al., 1980], [Rossolenko et al., 2001], [Yang et al., 2006], the melt column height in this case is $H_e^1 = \dfrac{1}{\rho \cdot g}\left(2198 - p_g^e\right)$, where $p_g^e \geq 0$ is the pressure of the gas flow (introduced in the furnace for release, the heat from the outer wall of the tube). When $p_g^e = 0$, then $H_e^1 = 8.96 \times 10^{-2}$ [m] i.e. the shaper's outer top level has to be with $H_e^1 = 8.96 \times 10^{-2}$ [m] above the crucible melt level.

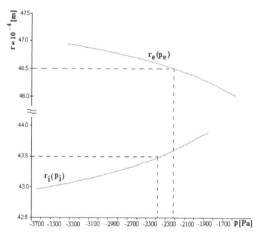

Fig. 4. The tube outer radius and inner radius versus p_e and p_i

To create a convex meniscus appropriate for the growth of a tube having the inner radius $r_i^1 = 4.35 \times 10^{-3}$ [m] ($m_1 = 1.03571$), according to the **Theorem 6** (Appendix 1), p_i has to be chosen in the range: $[-3723.32, -847.10]$ [Pa]. According to the **Corollary 8** (Appendix 1), from this range for the values of p_i smaller than -1945.80 [Pa], the point r_i where the condition $z_i'(r_i) = \tan\left(\frac{\pi}{2} - \alpha_g\right)$ is satisfied is close to 4.5×10^{-3} [m]. Therefore, we have to find now for $p_i = -1945.80$ [Pa] the point r_i where the above condition is achieved. This can be made by integrating numerically the system:

$$\begin{cases} \dfrac{dz_i}{dr} = \tan\alpha_i \\ \dfrac{d\alpha_i}{dr} = \dfrac{\rho \cdot g \cdot z_i - p_i}{\gamma} \cdot \dfrac{1}{\cos\alpha_i} - \dfrac{1}{r} \cdot \tan\alpha_i \end{cases} \tag{11}$$

for $z(R_{gi}) = 0$, $\alpha(R_{gi}) = \alpha_c$ and $p_i = -1945.80$ [Pa]. (see Fig. 3).

Since the obtained r_i is $r_i = 4.390 \times 10^{-3}$ [m] and it is higher than the desired value $r_i^1 = 4.35 \times 10^{-3}$ [m], we have to choose the value of p_i in the range $[-3723.32 - 1945.80]$ [Pa].

The results of the integrations of the system (11) for $z(R_{gi}) = 0$, $\alpha(R_{gi}) = \alpha_c$ and for different values of p_i in this range are represented in Fig. 4.

This figure shows that the inner radius $r_i^1 = 4.35 \times 10^{-3}$ [m] is obtained for $p_i^1 = -2434$ [Pa].

Taking $p_m \approx 0$ [Eriss et al., 1980], [Rossolenko et al., 2001], [Yang et al., 2006], the melt column height is $H_i^1 = \dfrac{1}{\rho \cdot g}\left(2434 - p_g^i\right)$, where $p_g^i \geq 0$ is the pressure of the gas flow (introduced in the furnace for releasing the heat from the inner wall of the tube). When $p_g^i = 0$, then $H_i^1 = 9.92 \times 10^{-2}$ [m], i.e. the shaper's inner top level has to be with $H_i^1 = 9.92 \times 10^{-2}$ [m] above the crucible melt level. When $p_g^i \geq 2434$, then H_i^1 is negative, i.e. the crucible melt level has to be above the shaper's inner top level.

To create a convex meniscus appropriate for the growth of a tube with the outer radius $r_e^1 = 4.65 \times 10^{-3}$ [m] and inner radius $r_i^1 = 4.35 \times 10^{-3}$ [m], when the shaper's inner top is at the same level as the shaper's outer top, we have to take: $\dfrac{1}{\rho \cdot g}\left(2198 - p_g^e\right) = \dfrac{1}{\rho \cdot g}\left(2434 - p_g^i\right)$.

It follows that the pressure of the gas flow, introduced in the furnace for releasing the heat from the inner wall of the tube has to be higher than the pressure of the gas flow, introduced in the furnace for releasing the heat from the outer wall of the tube and we have to take: $p_g^i - p_g^e = 236$ [Pa].

3.2 Concave free surface creation

In this section, it will be shown in which way the inequalities presented in Appendix 2 can be used for the creation of an appropriate static concave meniscus by the choice of p_e and p_i [Balint&Balint, 2009a].

Inequalities (A.2.1) establish the range in which the pressure difference p_e has to be chosen in order to obtain a static meniscus with concave outer free surface appropriate for the growth of a tube of outer radius equal to $\dfrac{R_{ge}}{n}$.

If the pressure difference satisfies inequality (A.2.2) then a static meniscus with concave outer free surface is obtained which is appropriate for the growth of a tube of outer radius in the range $\left[\dfrac{R_{ge}}{n}, \dfrac{R_{ge}}{n'}\right]$.

Theorem 13 (Appendix 2) shows that a static meniscus having a concave outer free surface appropriate for the growth of a tube of outer radius $r_e \in \left[\dfrac{R_{gi} + R_{ge}}{2}, R_{ge}\right]$ is stable.

Inequalities (A.2.3) establish the range in which the pressure difference p_i has to be chosen in order to obtain a static meniscus with convex inner free surface appropriate for the growth of a tube of inner radius equal to $m \cdot R_{gi}$.

If the pressure difference p_i satisfies inequality (A.2.4) then a static meniscus with concave inner free surface is obtained which is appropriate for the growth of a tube of inner radius in the range $\left[m' \cdot R_{gi}, m \cdot R_{gi}\right]$.

Theorem 16 (Appendix 2) shows that a static meniscus having a concave inner free surface appropriate for the growth of a tube of inner radius $r_i \in \left[R_{gi}, \dfrac{R_{gi} + R_{ge}}{2}\right]$ is stable.

Computations were performed for an InSb tube growth: $\alpha_c = 63.8^0$; $\alpha_g = 28.9^0$; $\rho = 6582 \left[kg / m^3\right]$; $\gamma = 4.2 \times 10^{-1}[N / m]$.

If there exists a concave outer free surface, appropriate for the growth of a tube of outer radius $r_e^1 = 4.65 \times 10^{-3}[m]$ ($n_1 = 1.03226$), then according to the **Theorem 11** (Appendix 2) this can be obtained for a value of p_e in the range $(134.85; 164.49)[Pa]$.

Taking into account the above fact, in order to create a concave outer free surface, appropriate for the growth of a tube of which outer radius is equal to $r_e^1 = 4.65 \times 10^{-3}[m]$ we have solved the i.v.p. (A.1.3) for different values of p_e in the range $(134.85; 164.49)[Pa]$.

More precisely, we have integrated the system (10) for $z_e(R_{ge}) = 0$, $z_e{'}(R_{ge}) = -\tan \alpha_c$ and different p_e. The obtained outer radii r_e versus p_e are represented in Fig.5, which shows that the desired outer radius $r_e^1 = 4.65 \times 10^{-3}[m]$ is obtained for $p_e{'} = 149.7[Pa]$.

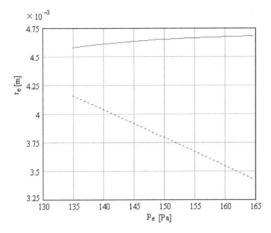

Fig. 5. Outer radii r_e versus p_e in the range $(134.85; 164.49)[Pa]$.

Actually, as it can be seen in the same figure, for $p_e' = 149.7[Pa]$ we can also obtain a second outer radius $r_e^2 = 3.8 \times 10^{-3}[m]$, which is not in the desired range $\left(\dfrac{R_{gi} + R_{ge}}{2}, R_{ge} \right)$.

Moreover, the outer free surface of this meniscus is not globally concave; it is a convex-concave meniscus (Fig.6).

Taking into account $p_m \approx 0$ [Eriss, 1980], [Rossolenko, 2001], [Yang, 2006], the melt column height in this case is $H_e' = -\dfrac{1}{\rho \cdot g} \cdot [p_e' + p_g^e]$, where $p_g^e \geq 0$ is the pressure of the gas flow (introduced in the furnace for releasing the heat from the outer side of the tube wall). When $p_g^e = 0$, then H_e' is negative, $H_e' = -2.31 \times 10^{-3}[m]$, i.e. the crucible melt level has to be with $-H_e' = 2.31 \times 10^{-3}[m]$ above the shaper top level.

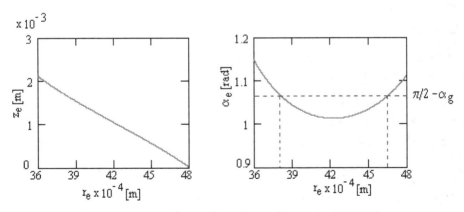

Fig. 6. Non globally concave outer free surface obtained for $p_e' = 149.7[Pa]$.

If there exists a concave inner free surface for the growth of a tube of inner radius

$r_i' = 4.35 \times 10^{-3} [m]$ ($m_1 = 1.03571$), then according to the **Theorem 14** (Appendix 2), this can be obtained for a value of p_i which is in the range $(-31.46, -1.07)[Pa]$.

Taking into account the above fact, in order to create a concave inner free surface, appropriate for the growth of a tube whose inner radius is equal to $r_i' = 4.35 \times 10^{-3} [m]$, we have solved the i.v.p. (A.1.8) for different values of p_i in the range $(-31.46, -1.07)[Pa]$. More precisely, we have integrated the system (11) for $z_i(R_{gi}) = 0$, $z_i'(R_{gi}) = \tan \alpha_c$ and different p_i. The obtained inner radii r_i versus p_i are represented in Fig.7 which shows that the desired inner radius $r_i^1 = 4.35 \times 10^{-3} [m]$ is obtained for $p_i' = -16.2[Pa]$.

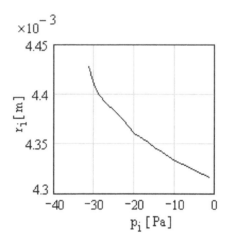

Fig. 7. Inner radii r_i versus p_i in the range $(-31.46, -1.07)[Pa]$.

Taking $p_m \approx 0$ [Eriss et al., 1980], [Rossolenko et al., 2001], [Yang et al. 2006], the melt column height in this case is $H_i' = -\dfrac{1}{\rho \cdot g} \cdot [p_i' + p_g^i]$, where $p_g^i \geq 0$ is the pressure of the gas flow (introduced in the furnace for releasing the heat from the inner side of the tube wall). When $p_g^i = 0$, then H_i' is positive, $H_i' = 0.25 \times 10^{-3} [m]$, i.e. the crucible melt level has to be with $H_i' = 0.25 \times 10^{-3} [m]$ under the shaper top level. To create a concave meniscus, appropriate for the growth of a tube with outer radius $r_e' = 4.65 \times 10^{-3}$ [m] and inner radius $r_i' = 4.35 \times 10^{-3}$ [m] the melt column heights (with respect to the crucible melt level) have to be $H_i^1 = -\dfrac{1}{\rho \cdot g} \cdot \left[-16.2 + p_g^i\right]$ and $H_e^1 = -\dfrac{1}{\rho \cdot g} \cdot \left[149.7 + p_g^e\right]$. When the shaper's outer top is at the same level as the shaper's inner top, with respect to the crucible melt level, then the relation $H_e^1 = H_i^1$ holds . It follows that the pressure of the gas flow, introduced in the furnace for releasing the heat from the inner wall of the tube, p_g^i has to be higher than the pressure of the gas flow introduced in the furnace for releasing the heat from the outer wall of the tube, p_g^e; $p_g^i - p_g^e = 149.7 + 16.2 = 165.9 [Pa]$.

4. The angles $\bar{\alpha}_e(r_e, h_e, p_e)$ and $\bar{\alpha}_i(r_i, h_i, p_i)$ which appear in system (1) describing the dynamics of the outer and inner radius of a tube, grown by the E.F.G. technique

4.1 The procedure for the determination of the angles $\bar{\alpha}_e(r_e, h_e, p_e)$ and $\bar{\alpha}_i(r_i, h_i, p_i)$ for a convex free surface

The angles $\bar{\alpha}_e(r_e, h_e, p_e)$ ($\bar{\alpha}_i(r_i, h_i, p_i)$) represent the angles between the tangent line to the outer free surface (inner free surface) of the meniscus at the three phase point, of coordinates (r_e, h_e) ((r_i, h_i)) and the horizontal axis Or. These angles can fluctuate during the growth. The deviation of the tangent to the crystal outer (inner) free surface at the triple point from the vertical is the difference $\bar{\alpha}_e(r_e, h_e, p_e) - \left(\dfrac{\pi}{2} - \alpha_g\right)$, $\left(\bar{\alpha}_i(r_i, h_i, p_i) - \left(\dfrac{\pi}{2} - \alpha_g\right)\right)$ (Fig. 2), where α_g is the growth angle. The deviation can fluctuate also and the outer (inner) radius r_e (r_i) is constant when the deviation is constant equal to zero

The angles $\bar{\alpha}_e(r_e, h_e, p_e)$ and $\bar{\alpha}_i(r_i, h_i, p_i)$ cannot be obtained directly from the Young-Laplace equation. For this reason for this equation the following strategy is adopted: two conditions are imposed at the outer radius R_{ge} of the shaper (inner radius R_{gi} of the shaper) $z_e(R_{ge}) = 0$; $z_e'(R_{ge}) = -\tan(\alpha_c^e)$ ($z_i(R_{gi}) = 0$, $z_i'(R_{gi}) = \tan(\alpha_c^i)$). In the last condition α_c^e (α_c^i) is a parameter which can fluctuate in a certain range, during the growth. For different values of α_c^e (α_c^i) in a given range the solution $z_e(r; \alpha_c^e, p_e)$ ($z_i(r; \alpha_c^i, p_i)$) of the Young-Laplace equation which satisfies the conditions $z_e(R_{ge}) = 0$; $z_e'(R_{ge}) = -\tan(\alpha_c^e)$ at R_{ge} ($(z_i(R_{gi}) = 0$, $z_i'(R_{gi}) = \tan(\alpha_c^i))$ at R_{gi}) is found.

With $z_e(r; \alpha_c^e, p_e)$ ($z_i(r; \alpha_c^i, p_i)$) the function $\alpha_e(r_e; \alpha_c^e, p_e) = -\arctan z_e'(r_e; \alpha_c^e, p_e)$ ($\alpha_i(r_i; \alpha_c^i, p_i) = \arctan z_i'(r_i; \alpha_c^i, p_i)$) is constructed. After that, from $h_e = z_e(r; \alpha_c^e, p_e)$ ($h_i = z_i(r; \alpha_c^i, p_i)$) α_c^e (α_c^i) is expressed as function of r_e, h_e and p_e (r_i, h_i and p_i).

$\alpha_c^e = \alpha_c^e(r_e; h_e, p_e)$ ($\alpha_c^i = \alpha_c^i(r_i; h_e, p_i)$) is introduced in $\alpha_e(r_e; \alpha_c^e, p_e)$ ($\alpha_i(r_i; \alpha_c^i, p_i)$) obtaining the function

$$\bar{\alpha}_e(r_e, h_e, p_e) = \alpha_e(r_e; \alpha_c^e(r_e; h_e, p_e), p_e) \qquad (\bar{\alpha}_i(r_i, h_i, p_i) = \alpha_i(r_i; \alpha_c^i(r_i; h_i, p_i), p_i)).$$

To the best of our knowledge, there is no algorithm in the literature concerning the construction of $\bar{\alpha}_e(r_e, h_e, p_e)$ ($\bar{\alpha}_i(r_i, h_i, p_i)$) at the level of generality presented here.

Due to the nonlinearity, the above described procedure can't be realized analytically. This is the reason why for the construction of the function $\bar{\alpha}_e(r_e;\alpha_c,p_e)$ in [Balint&Tanasie, 2010] the following numerical procedure was conceived:

Step 1. For a given α_c^{e0}; $\alpha_c^{e0} \in \left(0, \frac{\pi}{2} - \alpha_g\right)$ and $n = \dfrac{2 \cdot R_{ge}}{R_{ge} + R_{gi}}$ an $n' \in (1,n)$ is found such

that $E_1^e\left(n',\alpha_c^{e0}\right) < E_2^e\left(n,\alpha_c^{e0}\right)$ where:

$$E_1^e\left(n',\alpha_c^{e0}\right) = -\gamma \cdot \frac{\frac{\pi}{2} - \left(\alpha_c^{e0} + \alpha_g\right)}{R_{ge}} \cdot \frac{n'}{n'-1} \cdot \sin\alpha_g + \rho_1 \cdot g \cdot R_{ge} \cdot \frac{n'-1}{n'} \cdot \tan\left(\frac{\pi}{2} - \alpha_g\right) +$$

$$+ \frac{\gamma}{R_{ge}} \cdot n' \cdot \cos\alpha_g \tag{12}$$

$$E_2^e(n,\alpha_c^{e0}) = -\gamma \cdot \frac{\frac{\pi}{2} - \left(\alpha_c^{e0} + \alpha_g\right)}{R_{ge}} \cdot \frac{n}{n-1} \cdot \cos\alpha_c^{e0} + \frac{\gamma}{R_{ge}} \cdot \sin\alpha_c^{e0}$$

Step 2. For α_c^e a range $\left[\underline{\alpha}_c^e, \overline{\alpha}_c^e\right]$ is determined such that $0 < \underline{\alpha}_c^e < \alpha_c^{e0} < \overline{\alpha}_c^e < \frac{\pi}{2} - \alpha_g$ and for

every $\alpha_c^e \in \left[\underline{\alpha}_c^e, \overline{\alpha}_c^e\right]$ the inequality $E_1^e\left(n',\alpha_c^e\right) < E_2^e\left(n,\alpha_c^e\right)$ holds.

Step 3. For p_e the range $\left[\underline{p}_e, \overline{p}_e\right]$ defined by:

$$\underline{p}_e = \sup_{\alpha_c^e \in \left[\underline{\alpha}_c^e, \overline{\alpha}_c^e\right]} E_1^e\left(n',\alpha_c^e\right) \qquad \overline{p}_e = \inf_{\alpha_c^e \in \left[\underline{\alpha}_c^e, \overline{\alpha}_c^e\right]} E_2^e\left(n,\alpha_c^e\right) \tag{13}$$

is considered.

Step 4. In the range $\left[\underline{\alpha}_c^e, \overline{\alpha}_c^e\right]$ a set of l different values of α_c^e is chosen.

Step 5. In the range $\left[\underline{p}_e, \overline{p}_e\right]$ a set of m different values of p_e is chosen.

Step 6. In a given range $\left[\underline{\alpha}_e, \overline{\alpha}_e\right]$ possessing the property $\overline{\alpha}_c^e < \underline{\alpha}_e < \frac{\pi}{2} - \alpha_g < \overline{\alpha}_e < \frac{\pi}{2}$ a set of j values of α_e is chosen: $\underline{\alpha}_e = \alpha_e^1 < \alpha_e^2 < ... < \alpha_e^j = \overline{\alpha}_e$.

Step 7. For a given p_e^k, $k = \overline{1,m}$ and α_c^{eq}, $q = \overline{1,l}$ the solution of the system (10) corresponding to the conditions: $z_e(R_{ge}) = 0$, $\alpha_e(R_{ge}) = \alpha_c^{eq}$ is determined numerically obtaining the functions (profiles curves Refs [Tatarchenko, 1993]): $z_e = z_e(r;\alpha_c^{eq},p_e^k)$ and $\alpha_e = \alpha_e(r;\alpha_c^{eq},p_e^k)$.

Step 8. The values r_{kqs}^e for which $\alpha_e(r_{kqs}^e;\alpha_c^{eq},p_e^k) = \alpha_e^s \in \left[\underline{\alpha}_e, \overline{\alpha}_e\right]$, $k = \overline{1,m}$, $q = \overline{1,l}$ are determined.

Step 9. The values $h^e_{kqs} = z_e(r^e_{kqs}; \alpha^{eq}_c, p^k_e)$ are found.

Step 10. Fitting the data r^e_{kqs} h^e_{kqs} p^k_e and α^s_e, the function $\bar{\alpha}_e(r_e, h_e, p_e)$ is found.

For the same reason as in the case of $\bar{\alpha}_e(r_e, h_e, p_e)$ for the construction of $\bar{\alpha}_i = \bar{\alpha}_i(r_i, h_i, p_i)$ the following numerical procedure was conceived:

Step 1. For $\alpha^{i0}_c = \alpha^{e0}_c$ and $m \in \left(1, \dfrac{R_{ge} + R_{gi}}{2 \cdot R_{gi}}\right]$ an $m' \in (1, m)$ is determined such that

$E^i_1\left(m', \alpha^{i0}_c\right) < E^i_2\left(m, \alpha^{i0}_c\right)$ where:

$$E^i_1\left(m', \alpha^{i0}_c\right) = -\gamma \cdot \frac{\pi/2 - \left(\alpha^{i0}_c + \alpha_g\right)}{(m' - 1) \cdot R_{gi}} \cdot \sin \alpha_g + \rho_1 \cdot g \cdot R_{gi} \cdot (m' - 1) \cdot \tan\left(\pi/2 - \alpha_g\right) -$$

$$- \frac{\gamma}{m' \cdot R_{gi}} \cdot \sin \alpha^{i0}_c \tag{14}$$

$$E^i_2(m, \alpha^{i0}_c) = -\gamma \cdot \frac{\pi/2 - \left(\alpha^{i0}_c + \alpha_g\right)}{(m - 1) \cdot R_{gi}} \cdot \cos \alpha^{i0}_c - \frac{\gamma}{R_{gi}} \cdot \cos \alpha_g$$

Step 2. For α^i_c a range $\left[\underline{\alpha}^i_c, \overline{\alpha}^i_c\right]$ is determined such that $0 < \underline{\alpha}^i_c < \alpha^i_c < \overline{\alpha}^i_c < \pi/2 - \alpha_g$ and for every $\alpha^i_c \in \left[\underline{\alpha}^i_c, \overline{\alpha}^i_c\right]$, the inequality $E^i_1\left(m', \alpha^i_c\right) < E^i_2\left(m, \alpha^i_c\right)$ holds.

Step 3. For p_i the range $\left[\underline{p}_i, \overline{p}_i\right]$ defined by:

$$\underline{p}_i = \sup_{\alpha^i_c \in \left[\underline{\alpha}^i_c, \overline{\alpha}^i_c\right]} E^i_1\left(m', \alpha^i_c\right) \qquad \overline{p}_i = \inf_{\alpha^i_c \in \left[\underline{\alpha}^i_c, \overline{\alpha}^i_c\right]} E^i_2\left(m, \alpha^i_c\right) \tag{15}$$

is considered.

Step 4. In the range $\left[\underline{\alpha}^i_c, \overline{\alpha}^i_c\right]$ a set of l different values of α^i_c are chosen.

Step 5. In the range $\left[\underline{p}_i, \overline{p}_i\right]$ a set of n different values of p_i are chosen.

Step 6. In a given range $\left[\underline{\alpha}_i, \overline{\alpha}_i\right]$, possessing the property $\overline{\alpha}^i_c < \underline{\alpha}_i < \pi/2 - \alpha_g < \overline{\alpha}_i < \pi/2$, a set of j values of α_i are chosen: $\underline{\alpha}_i = \alpha^1_i < \alpha^2_i < ... < \alpha^j_i = \overline{\alpha}_i$.

Step 7. For a given p^k_i, $k = \overline{1, n}$ and α^{iq}_c, $q = \overline{1, l}$ the solution of the system (11) which satisfies the conditions: $z_i(R_{gi}) = 0$, $\alpha_i(R_{gi}) = \alpha^{iq}_c$ is found numerically obtaining the functions (profiles curves Ref. [Tatarchenko, 1993]): $z_i = z_i(r; \alpha^{iq}_c, p^k_i)$ and $\alpha_i = \alpha_i(r; \alpha^{iq}_c, p^k_i)$.

Step 8. The values r_{kqs}^i for which $\alpha_i(r_{kqs}^i;\alpha_c^{iq},p_i^k)=\alpha_i^s\in\left[\underline{\alpha}_e,\overline{\alpha}_e\right]$, $k=\overline{1,n}$, $q=\overline{1,l}$ are determined.

Step 9. The values $h_{kqs}^i = z_i(r_{kqs}^i;\alpha_c^{iq},p_i^k)$ are found.

Step 10. The function $\overline{\alpha}_i(r_i,h_i,p_i)$ is found by fitting the data r_{kqs}^i h_{kqs}^i p_i^k and α_i^s.

For the case of a silicon tube and the outer free surface the function:

$$\overline{\alpha}_e(r_e,h_e,p_e) = \frac{a_1(p_e)+a_2(p_e)\cdot r_e+a_3(p_e)\cdot r_e^2+a_4(p_e)\cdot h_e+a_5(p_e)\cdot h_e^2+a_6(p_e)\cdot h_e^3}{1+a_7(p_e)\cdot r_e+a_8(p_e)\cdot h_e+a_9(p_e)\cdot h_e^2+a_{10}(p_e)\cdot h_e^3}$$

with: $a_1(p_e) = 0.400314546-0.00494101\cdot p_e$ $\qquad a_2(p_e) = -374.789455+2.353132518\cdot p_e$

$a_3(p_e) = 56161.27414-244.013259\cdot p_e$ $\qquad a_4(p_e) = 393.2446055+0.010192507\cdot p_e$

$a_5(p_e) = 91143.39025-555.405995\cdot p_e$ $\qquad a_6(p_e) = -1097500000-1168100\cdot p_e$

$a_7(p_e) = -200.489376+0.001998751\cdot p_e$ $\qquad a_8(p_e) = 114.0391478+0.002922594\cdot p_e$

$a_9(p_e) = 83485.77784-477.770821\cdot p_e$ $\qquad a_{10}(p_e) = 415910000+476372.6048\cdot p_e$

was obtained and for the inner free surface the function:

$$\overline{\alpha}_i(r_i;h_i,p_i) = \frac{b_1(p_i)+b_2(p_i)\cdot r_i+b_3(p_i)\cdot r_i^2+b_4(p_i)\cdot h_i+b_5(p_i)\cdot h_i^2+b_6(p_i)\cdot h_i^3}{1+b_7(p_i)\cdot r_i+b_8(p_i)\cdot h_i+b_9(p_i)\cdot h_i^2+b_{10}(p_i)\cdot h_i^3}$$

with $\quad b_1(p_i) = 0.102617985+0.004931999\cdot p_i$ $\qquad b_2(p_i) = -172.7251-2.34898717\cdot p_i$

$b_3(p_i) = 29313.52855+277.5275727\cdot p_i$ $\qquad b_4(p_i) = 444.327355+0.367481455\cdot p_i$

$b_5(p_i) = -2687700-817.408936\cdot p_i$ $\qquad b_6(p_i) = 5760710000+3810000\cdot p_i$

$b_7(p_i) = -263.39243-0.00872161\cdot p_i$ $\qquad b_8(p_i) = 707.770183+0.334375383\cdot p_i$

$b_9(p_i) = -2709500-696.727273\cdot p_i$ $\qquad b_{10}(p_i) = 3090180000+1348660\cdot p_i$

was obtained. For $\quad p_e =-2000[Pa]$ and $p_i =-2242[Pa]$ the functions $\overline{\alpha}_e(r_e,h_e,p_e)$ and $\overline{\alpha}_i(r_i,h_i,p_i)$ are represented in Fig. 8.:

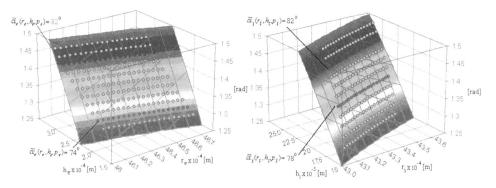

Fig. 8. The graphics of $\overline{\alpha}_e(r_e,h_e,p_e)$ ($p_e =-2000\,[Pa]$) and $\overline{\alpha}(r_i,h_i,p_i)$ ($p_i =-2242\,[Pa]$)

4.2 The procedure for the determination of the angles $\bar{\alpha}_e(r_e,h_e,p_e)$ and $\bar{\alpha}(r_i,h_i,p_i)$ for a concave free surface

The numerical procedure for the construction of the function $\bar{\alpha}_e(r_e,h_e,p_e)$ for a concave free surface (when $\alpha_c > \dfrac{\pi}{2} - \alpha_g$) is similar to those applied for a convex free surface. Only **Step 1-Step 3.** present the some differences. For the outer free surface we have to consider:

Step 1. For a given α_c^{e0}; $\alpha_c^{e0} \in \left(\dfrac{\pi}{2} - \alpha_g, \dfrac{\pi}{2}\right)$ and $n = \dfrac{2 \cdot R_{ge}}{R_{ge} + R_{gi}}$ an $n' \in (1,n)$ is found such that $E_1^e\left(n,\alpha_c^{e0}\right) < E_2^e\left(n',\alpha_c^{e0}\right)$ where:

$$E_1^e\left(n,\alpha_c^{e0}\right) = \gamma \cdot \frac{\alpha_c^{e0} + \alpha_g - \pi/2}{R_{ge}} \cdot \frac{n}{n-1} \cdot \sin\alpha_g + \rho_1 \cdot g \cdot R_{ge} \cdot \frac{n-1}{n} \cdot \tan\alpha_c^{e0} + \frac{\gamma}{R_{ge}} \cdot n \cdot \sin\alpha_c^{e0}$$

$$E_2^e(n',\alpha_c^{e0}) = \gamma \cdot \frac{\alpha_c^{e0} + \alpha_g - \pi/2}{R_{ge}} \cdot \frac{n'}{n'-1} \cdot \cos\alpha_c^{e0} + \frac{\gamma}{R_{ge}} \cdot \cos\alpha_g \tag{16}$$

Step 2. For α_c^e a range $\left[\underline{\alpha}_c^e, \overline{\alpha}_c^e\right]$ is determined such that $\pi/2 - \alpha_g < \underline{\alpha}_c^e < \alpha_c^{e0} < \overline{\alpha}_c^e < \pi/2$ and the inequality $\displaystyle\sup_{\alpha_c^e \in \left[\underline{\alpha}_c^e, \overline{\alpha}_c^e\right]} E_1^e\left(n,\alpha_c^e\right) < \inf_{\alpha_c^e \in \left[\underline{\alpha}_c^e, \overline{\alpha}_c^e\right]} E_2^e\left(n',\alpha_c^e\right)$ holds.

Step 3. For p_e the range $\left[\underline{p}_e, \overline{p}_e\right]$ defined by:

$$\underline{p}_e = \sup_{\alpha_c^e \in \left[\underline{\alpha}_c^e, \overline{\alpha}_c^e\right]} E_1^e\left(n,\alpha_c^e\right) \qquad \overline{p}_e = \inf_{\alpha_c^e \in \left[\underline{\alpha}_c^e, \overline{\alpha}_c^e\right]} E_2^e\left(n',\alpha_c^e\right) \tag{17}$$

is considered.

For the inner free surface we have to make:

Step 1. For $\alpha_c^{i0} = \alpha_c^{e0}$ and $m \in \left(1, \dfrac{R_{ge} + R_{gi}}{2 \cdot R_{gi}}\right]$ an $m' \in (1,m)$ is determined such that $E_1^i\left(m,\alpha_c^{i0}\right) < E_2^i\left(m',\alpha_c^{i0}\right)$ where:

$$E_1^i\left(m,\alpha_c^{i0}\right) = \gamma \cdot \frac{\alpha_c^{i0} + \alpha_g - \pi/2}{(m-1) \cdot R_{ge}} \cdot \sin\alpha_g + \rho_1 \cdot g \cdot R_{gi} \cdot (m-1) \cdot \tan\alpha_c^{i0} - \frac{\gamma}{m \cdot R_{gi}} \cdot \cos\alpha_g$$

$$E_2^i(m',\alpha_c^{i0}) = \gamma \cdot \frac{\alpha_c^{i0} + \alpha_g - \pi/2}{(m'-1) \cdot R_{gi}} \cdot \cos\alpha_c^{i0} + \frac{\gamma}{R_{gi}} \cdot \sin\alpha_c^{i0} \tag{18}$$

Step 2. For α_c^i a range $\left[\underline{\alpha}_c^i, \overline{\alpha}_c^i\right]$ is determined such that $\pi/2 - \alpha_g < \underline{\alpha}_c^i < \alpha_c^{i0} < \overline{\alpha}_c^i < \pi/2$ and the inequality $\displaystyle\sup_{\alpha_c^i \in \left[\underline{\alpha}_c^i, \overline{\alpha}_c^i\right]} E_1^i\left(m,\alpha_c^i\right) < \inf_{\alpha_c^i \in \left[\underline{\alpha}_c^i, \overline{\alpha}_c^i\right]} E_2^i\left(m',\alpha_c^i\right)$ holds.

Step 3. For p_i the range $\left[\underline{p}_i, \overline{p}_i\right]$ defined by:

$$\underline{p}_i = \sup_{\alpha_c^i \in \left[\underline{\alpha}_c^i, \overline{\alpha}_c^i\right]} E_1^i\left(m, \alpha_c^i\right) \qquad \overline{p}_i = \inf_{\alpha_c^i \in \left[\underline{\alpha}_c^i, \overline{\alpha}_c^i\right]} E_2^i\left(m', \alpha_c^i\right) \qquad (19)$$

is considered.

In the case of of the InSb tube considered in section 2.2, for the outer free surface the function:

$$\overline{\alpha}_e\left(r_e, h_e, p_e\right) = \frac{a_1(p_e) + a_2\left(p_e\right)\cdot r_e + a_3\left(p_e\right)\cdot r_e^2 + a_4\left(p_e\right)\cdot h_e + a_5\left(p_e\right)\cdot h_e^2 + a_6\left(p_e\right)\cdot h_e^3}{1 + a_7\left(p_e\right)\cdot r_e + a_8\left(p_e\right)\cdot h_e + a_9\left(p_e\right)\cdot h_e^2 + a_{10}\left(p_e\right)\cdot h_e^3}$$

with

$a_1(p_e) = 16.95005004 - 0.1803412 \cdot p_e$ $\qquad$ $a_2(p_e) = \text{-}6991.91018 + 76.11771837 \cdot p_e$

$a_3(p_e) = 721251.8226 - 8031.90955 \cdot p_e$ $\qquad$ $a_4(p_e) = 82.08229042 - 1.41357516 \cdot p_e$

$a_5(p_e) = -820223.151 + 11947.58366 \cdot p_e$ $\qquad$ $a_6(p_e) = 75949100 - 3370800 \cdot p_e$

$a_7(p_e) = -207.085329 - 0.00600066 \cdot p_e$ $\qquad$ $a_8(p_e) = 0.10447655 - 0.82800845 \cdot p_e$

$a_9(p_e) = -544970.373 + 8774.44 \cdot p_e$ $\qquad$ $a_{10}(p_e) = -237870000 - 663676.426 \cdot p_e$

was obtained. For the inner free surface the function:

$$\overline{\alpha}_i\left(r_i; h_i, p_i\right) = \frac{b_1(p_i) + b_2\left(p_i\right)\cdot r_i + b_3\left(p_i\right)\cdot h_i + b_4\left(p_i\right)\cdot h_i^2 + b_5\left(p_i\right)\cdot h_i^3}{1 + b_6\left(p_i\right)\cdot r_i + b_7\left(p_i\right)\cdot r_i^2 + b_8\left(p_i\right)\cdot h_i + b_9\left(p_i\right)\cdot h_i^2}$$

with $\quad b_1(p_i) = \text{-}0.03323515 + 0.0000618675 \cdot \text{p}_\text{i}$ $\qquad$ $b_2(p_i) = 7.907785322 \text{-} 0.01471419 \cdot \text{p}_\text{i}$

$b_3(p_i) = \text{-}10.9920301 + 0.019065783 \cdot \text{p}_\text{i}$ $\qquad$ $b_4(p_i) = 81321.53642 + 10.42303935 \cdot \text{p}_\text{i}$

$b_5(p_i) = \text{-}6164200 \text{-} 109065.464 \cdot \text{p}_\text{i}$ $\qquad$ $b_6(p_i) = \text{-}478.279159 + 0.004407571 \cdot \text{p}_\text{i}$

$b_7(p_i) = 57185.4894 \text{-} 1.04591888 \cdot \text{p}_\text{i}$ $\qquad$ $b_8(p_i) = \text{-}7.40209299 + 0.012781953 \cdot \text{p}_\text{i}$

$b_9(p_i) = 57826.05599 + 9.103117035 \cdot \text{p}_\text{i}$

was obtained. For $p_e = 180[Pa]$ and $p_i = 290[Pa]$ the functions $\overline{\alpha}_e(r_e, h_e, p_e)$ and $\overline{\alpha}_i(r_i, h_i, p_i)$ are represented in Fig. 9:

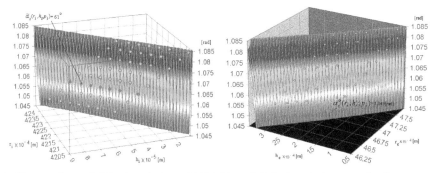

Fig. 9. The graphics of $\overline{\alpha}_i(r_i, h_i, p_i)$ ($p_i = 290[Pa]$) and $\overline{\alpha}_i(r_i, h_i, p_i)$ ($p_e = 180[Pa]$)

5. Setting the pulling rate, the thermal and capillary conditions

In this section it will be shown that the results presented in the above sections can be used for setting the pulling rate, the thermal and capillary conditions in view of an experiment [Tanasie&Balint, 2010].

According to [Tatarchenko, 1993] at the level of the crystallization front h the crystallization rate v_c^j is given by:

$$v_c^j = \frac{1}{\Lambda \cdot \rho_1} \cdot \left[\lambda_1 \cdot G_1^j(r_e, r_i, h) - \lambda_2 \cdot G_2^j(r_e, r_i, h) \right], j = 1, 2, 3. \tag{20}$$

The difference between the pulling rate v and the crystallization rate v_c^j is equal to the crystallization front displacement rate $\dfrac{dh^j}{dt}$, $j = 1, 2, 3$.

In order to keep the crystallization front level h^j constant, the pulling rate and the thermal conditions have to satisfy the following conditions:

$$v - \frac{1}{\Lambda \cdot \rho_1} \cdot \left[\lambda_1 \cdot G_1^j(r_e, r_i, h) - \lambda_2 \cdot G_2^j(r_e, r_i, h) \right] = 0, j = 1, 2, 3 \tag{21}$$

When the radii r_e, r_i and the length L of the tube, which has to be grown, are prior given, and h is known, then the condition (21) can be regarded as an equation in which the pulling rate v is unknown. If this equation has a positive solution v, it depends on the following parameters: h, $T_{en}(0)$, T_0, and k. The setting of the pulling rate, thermal conditions means the choice of v, $T_{en}(0)$, T_0 and k such that the following conditions be satisfied:

- $300 < T_{en}(0) < T_m < T_0$; $0 < k < \dfrac{T_{en}(0) - 300}{L}$;
- equation (21) has a positive solution v in an acceptable range.
- v is practically the same for every L': $L_0 \le L' \le L$ (L_0 = the seed length).

The setting of the capillary condition means to take the tube radii r_e, r_i (prior given) and the crystallization front level h_c, determined form (21) (for the above chosen v, $T_{en}(0)$, T_0, k) and find the pressures p_e, p_i solving the followings equations:

$$\overline{\alpha}_e(r_e, h_c, p_e) = \frac{\pi}{2} - \alpha_g \text{ and } \overline{\alpha}_i(r_i, h_c, p_i) = \frac{\pi}{2} - \alpha_g \tag{22}$$

If the solutions p_e, p_i of this equations are in the range for which $\overline{\alpha}_e$, $\overline{\alpha}_i$ was build up, then the values p_e, p_i will be used to set p_g^e, p_g^i, H_e, H_i using (2) with $p_m = 0$ or

$$H_e = -\frac{p_g^e + p_e}{\rho_1 \cdot g}, \ H_i = -\frac{p_g^i + p_i}{\rho_1 \cdot g} \tag{23}$$

For the growth of a silicon tube with convex profile curves the following numerical data will be used: $\rho_1 = 2.5 \times 10^3$ [kg/m³]; $\rho_2 = 2.3 \times 10^3$ [kg/m³]; $T_m = 1683$ [K]; $\lambda_1 = 60$ [W/m · K]; $\lambda_2 = 21.6$ [W/m · K]; $\Lambda = 1.81 \times 10^6$ [J/kg]; $\chi_1 = \dfrac{\lambda_1}{c_1 \cdot \rho_1}$; $\chi_2 = \dfrac{\lambda_2}{c_2 \cdot \rho_2}$; $c_1 = 913$ [J/kg · K]; $c_2 = 703$ [J/kg · K]; $\mu_1 = 7300.42$ [K]; $\mu_2 = 2822.58$ [K]; $R_{gi} = 4.2 \times 10^{-3}$ [m]; $R_{ge} = 4.8 \times 10^{-3}$ [m]; $R_i^c = 4.339 \times 10^{-3}$ [m]; $R_e^c = 4.66 \times 10^{-3}$ [m]; $L_1 = 0.4$ [m]; $L_2 = 0.2$ [m]; $L_3 = 0.1$ [m].

Step 1. A stable static outer meniscus is chosen, whose characteristic parameters r_e, h, p_e are in the range where $\overline{\alpha}_e(r_e, h, p_e)$ is valid and for which r_e is close to r_e^c. In the case considered here such a static meniscus is obtained for $p_e = -1980$ [Pa] and its characteristic parameters are: $r_e = 4.660112250074 \times 10^{-3}$ [m] and $h = 2.14370857185 \times 10^{-4}$ [m].

Step 2. An initial input for T_0, $T_{en}(0)$ and k has to be chosen. For T_0, the start can be $T_0 = T_m + 1$. Concerning $T_{en}(0)$ and k the start can be $T_{en}(0) = T_m - 1$ and $k = \dfrac{T_{en}(0) - 300}{L}$.

Using this input and the values r_e, h, r_i^c the value of the pulling rates $v_1, v_2, \ldots, v_{40}$ given by the equation (21) have to be found. If all these values are positive, then: if the average $\overline{v}$ and standard deviation σ of the set of values of v, are acceptable, then the average pulling rate $\overline{v}$ and the initial input thermal conditions can be set, else the initial input thermal conditions have to be reset lowering in general $T_{en}(0)$ and/or increasing T_0.

Step 3. Consider $\overline{v}$, $T_{en}(0)$, T_0, k obtained above and solve equation (22) for these values choosing $r_e = r_e^c$ and $r_i = r_i^c$ (the desired radii) and h unknown. Denote by h_c the obtained solution. Replace r_e^c, r_i^c, h_c in equation (23) and solve this equations finding p_e, p_i.

Step 4. Using p_e, p_i find $H_i - H_e$, for $p_g^e = p_g^i$ (in the case of an open crucible) or find $p_g^i - p_g^e$ for $H_i - H_e = 0$ (in the case of a closed crucible).

Following the above steps for the considered silicon tube growth, some of the computed possible settings, are presented in Table 1.

For the above settings the growth process stability analysis is made through the system of nonlinear ordinary differential equations (1) which governs the evolution of r_e, r_i, h for the established settings. It means to verify first of all that the desired r_e^c, r_i^c and the obtained h_c is a steady state of (1).

Furthermore to verify if at the start r_e^c, r_i^c, h_c are perturbed (i.e. the seed sizes are different from r_e^c, r_i^c) after a period of transition the values r_e^c, r_i^c, h_c are recovered. In other words, to verify if the steady state (r_e^c, r_i^c, h_c) is asymptotically stable. This last requirement is satisfied if the Hurwitz conditions are satisfied [Tatarchenko, 1993] i.e.:

$$- a_{11} - a_{22} - a_{33} > 0, \; - a_{11}a_{22}a_{33} + a_{31}a_{13}a_{22} > 0$$
$$(-a_{11} - a_{22} - a_{33})(-a_{31}a_{13} + a_{11}a_{22} + a_{22}a_{33} + a_{11}a_{33}) - (-a_{11}a_{22}a_{33} + a_{31}a_{13}a_{22}) > 0$$

(24)

$$a_{11} = -v \cdot \frac{\partial \overline{\alpha}_e(r_e^c, h_c, p_e)}{\partial R_e} \quad a_{12} = -v \cdot \frac{\partial \overline{\alpha}_e(r_e^c, h_c, p_e)}{\partial R_i} = 0 \quad a_{13} = -v \cdot \frac{\partial \overline{\alpha}_e(r_e^c, h_c, p_e)}{\partial h}$$

$$a_{21} = v \cdot \frac{\partial \overline{\alpha}_i(r_i^c, h_c, p_i)}{\partial R_e} = 0 \quad a_{22} = v \cdot \frac{\partial \overline{\alpha}_i(r_i^c, h_c, p_i)}{\partial R_i} \quad a_{23} = v \cdot \frac{\partial \overline{\alpha}_i(r_i^c, h_c, p_i)}{\partial h}$$

(25)

$$a_{31} = \frac{\partial S(r_e^c, r_i^c, h_c)}{\partial R_e} \quad a_{32} = \frac{\partial S(r_e^c, r_i^c, h_c)}{\partial R_i} \quad a_{33} = \frac{\partial S(r_e^c, r_i^c, h_c)}{\partial h}$$

$$S(r_e, r_i, h) = \overline{v} - \frac{1}{\Lambda \cdot \rho_1} \cdot [\lambda_1 \cdot G_1(r_e, r_i, h) - \lambda_2 \cdot G_2(r_e, r_i, h)]$$

	F₁	F₂	F₃		F₁	F₂	F₃
$\overline{v}$	8.32993·10	8.29566×10	8.29107 ×10	$H_i - H_e$	1.10328×10⁻²	1.10320×10⁻²	1.10324×10⁻²
T_0	1716.46	2110.76	2933.32	$p_g^i - p_g^e$	-270.58	-270.56	-270.57
$T_{en}(0)$	400.00	400.00	400.00	h_c	2.14380×10⁻⁴	2.14324×10⁻⁴	2.14371×10⁻⁴
k	250.00	500.00	1000.00	r_e^c	4.66000×10⁻³	4.65990×10⁻³	4.65990×10⁻³
p_e	-1981.44	-1982.93	-1981.61	r_i^c	4.33900×10⁻⁴	4.33900×10⁻⁴	4.33899×10⁻⁴
p_i	-2252.02	-2253.49	-2252.18				

Table 1. Possible settings of $\overline{v}$, T_0, $T_{en}(0)$, p_e, p_i, $p_g^i - p_g^e$

	a_{11}	a_{13}	a_{22}	a_{23}	a_{31}	a_{32}	a_{33}
F₁	-0.20278	-0.29136	-0.1984	0.285445	121.8176	0	-2534.0535
F₂	-0.20191	-0.29033	-0.19755	0.284434	27285.12	-27215.77	-42171.2316
F₃	-0.20183	-0.29002	-0.19747	0.284131	88368.31	-88368.31	-132203.96

Table 2. The coefficients a_{ij} of the linearized system in the steady states

The values of the numbers a_{ij} in the considered cases are given in Table 2. It is easy to verify that in all cases the Hurwitz condition are satisfied.

In Fig. 10 simulations of the silicon tube growth is presented when the seed length is 10^{-2} and the radii of the seed are $R_i = 4.3389 \times 10^{-3}$ [m]; $R_e = 4.659 \times 10^{-3}$ [m]. The meniscus height at the start is $h = 2.14838 \times 10^{-4}$ [m].

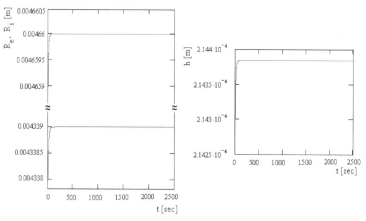

Fig. 10. The evolution of the outer radius, the inner radius of the tube and the meniscus height obtained integrating numerically the system (1) for $F_1 = 2 / R_e$,

$\bar{v} = 1.6585917365 \cdot 10^{-4}$ [m/s], $T_0 = 1714.81$ [K] and $T_{en}(0) = 400[K]$

6. Conclusions

Knowing the material constants (density, heat conductivity, etc), the size of the single crystal tube which will be grown from that material, the size of the shaper which will be used and the cooling gas temperature at the entrance, it is possible to predict values of pulling rate, temperature at the meniscus basis, cooling gas temperature at the exit, vertical temperature gradient in the furnace, inner and outer walls cooling gas pressure differences, melt column height differences, crystallization front level, which can be used for a stable growth.

According to the model the predicted values are not unique i.e. there are several possibility to obtain a tube with prior given size from a given material using the same shaper. So, even if in our computation the material and the size of the shaper and tube is the same as in [Eriss] experiment, the computed data given in Table 1 can be different from that used in the real experiment. For this reason our purpose is not rely to compare the computed results with the experimental data. Moreover we want to reveal that a tube of prior given size can be obtained by different settings and the model permit to compute such settings. The choice of a specific setting is the practical crystal grower decision. The model provide possible settings and can be helpful in a new experiment planning.

Concerning the limits of the model it is clear that it is limited in applicability, as all models. The main limits are those introduced by approximations made in equations defining the model.

7. Appendix 1. Inequalities for single crystal tube growth by E.F.G. technique - Convex outer and inner free surface

Consider the differential equation (6) for $\dfrac{R_{gi} + R_{ge}}{2} \le r_e \le R_{ge}$, α_c, α_g such that $\alpha_g \in \left(0, \dfrac{\pi}{2}\right)$, $0 < \alpha_c < \dfrac{\pi}{2} - \alpha_g$.

Definition 1. A solution $z = z(x)$ of the eq. (6) describes the outer free surface of a static meniscus on the interval $\left[r_e, R_{ge}\right]$ if it possesses the following properties: $z\left(R_{ge}\right) = 0$, $z'\left(r_e\right) = -\tan\left(\frac{\pi}{2} - \alpha_g\right)$, $z'\left(R_{ge}\right) = -\tan\alpha_c$ and $z(r)$ is strictly decreasing on $[r_e, R_{ge}]$. The described outer free surface is convex on $[r_e, R_{ge}]$ if $z''(r) > 0 \quad \forall r \in [r_e, R_{ge}]$.

Theorem 1. If there exists a solution of the eq. (6), which describes a convex outer free surface of a static meniscus on the closed interval $[r_e, R_{ge}]$, then for $n = \dfrac{R_{ge}}{r_e}$, p_e satisfy:

$$-\gamma \cdot \frac{\frac{\pi}{2} - \left(\alpha_c + \alpha_g\right)}{R_{ge}} \cdot \frac{n}{n-1} \cdot \cos\alpha_c + \frac{\gamma}{R_{ge}} \cdot \sin\alpha_c \le p_e$$

$$\le -\gamma \cdot \frac{\frac{\pi}{2} - \left(\alpha_c + \alpha_g\right)}{R_{ge}} \cdot \frac{n}{n-1} \cdot \sin\alpha_g + \frac{\rho \cdot g \cdot R_{ge} \cdot (n-1)}{n} \cdot \tan\left(\frac{\pi}{2} - \alpha_g\right) + \quad \text{(A.1.1)}$$

$$+ \frac{\gamma}{R_{ge}} \cdot n \cdot \cos\alpha_g$$

Theorem 2. Let be n such that $1 < n < \dfrac{2 \cdot R_{ge}}{R_{gi} + R_{ge}}$. If p_e satisfies the inequality:

$$p_e < -\gamma \cdot \frac{\frac{\pi}{2} - \left(\alpha_c + \alpha_g\right)}{R_{ge}} \cdot \frac{n}{n-1} \cdot \cos\alpha_c + \frac{\gamma}{R_{ge}} \cdot \sin\alpha_c \qquad \text{(A.1.2)}$$

then there exists $r_e \in \left[\dfrac{R_{ge}}{n}, R_{ge}\right]$ such that the solution of the initial value problem:

$$\begin{cases} z'' = \dfrac{\rho \cdot g \cdot z - p_e}{\gamma} \cdot \left[1 + (z')^2\right]^{\frac{3}{2}} - \dfrac{1}{r} \cdot \left[1 + (z')^2\right] \cdot z' \text{ for } \dfrac{R_{gi} + R_{ge}}{2} < r \le R_{ge} \\ z\left(R_{ge}\right) = 0, \ z'\left(R_{ge}\right) = -\tan\alpha_c \end{cases} \qquad \text{(A.1.3)}$$

on the interval $[r_e, R_{ge}]$ describes the convex outer free surface of a static meniscus.

Corollary 3. If for p_e the following inequality holds:

$$p_e < -2 \cdot \gamma \cdot \frac{\frac{\pi}{2} - \left(\alpha_c + \alpha_g\right)}{R_{ge} - R_{gi}} \cdot \cos\alpha_c + \frac{\gamma}{R_{ge}} \cdot \sin\alpha_c \qquad \text{(A.1.4)}$$

then there exists $r_e \in \left(\dfrac{R_{ge} + R_{gi}}{2}, R_{ge}\right)$ (close to $\dfrac{R_{ge} + R_{gi}}{2}$) such that the solution of the i.v.p. (12) on the interval $[r_e, R_{ge}]$ describes a convex outer free surface of a static meniscus.

Corollary 4. If for $1 < n' < n < \dfrac{2 \cdot R_{ge}}{R_{gi} + R_{ge}}$ the following inequalities holds:

$$-\gamma \cdot \frac{\pi/2 - (\alpha_c + \alpha_g)}{R_{ge}} \cdot \frac{n'}{n'-1} \cdot \sin\alpha_g + \rho \cdot g \cdot R_{ge} \cdot \frac{n'-1}{n'} \cdot \tan\left(\pi/2 - \alpha_g\right) + \frac{\gamma}{R_{ge}} \cdot n' \cos\alpha_g$$

$$< p_e < -\gamma \cdot \frac{\pi/2 - (\alpha_c + \alpha_g)}{R_{ge}} \cdot \frac{n}{n-1} \cdot \cos\alpha_c + \frac{\gamma}{R_{ge}} \cdot \sin\alpha_c$$

(A.1.5)

then there exists $r_e \in \left[\dfrac{R_{ge}}{n}, \dfrac{R_{ge}}{n'}\right]$ such that the solution of the i.v.p. (A.1.3) on the interval $[r_e, R_{ge}]$ describes a convex outer free surface of a static meniscus.

Theorem 5. If a solution $z_1 = z_1(r)$ of the eq. (6) describes a convex outer free surface of a static meniscus on the interval $[r_e, R_{ge}]$, then it is a weak minimum for the energy functional of the melt column (7).

Definition 2. A solution $z = z(x)$ of the eq.(8) describes the inner free surface of a static meniscus on the interval $\left[R_{gi}, r_i\right]$, $\left(R_{gi} < r_i < \dfrac{R_{gi} + R_{ge}}{2}\right)$ if it possesses the following properties: $z'(R_{gi}) = \tan\alpha_c$, $z'(r_i) = \tan\left(\pi/2 - \alpha_g\right)$, $z(R_{gi}) = 0$ and $z(r)$ is strictly increasing on $[R_{gi}, r_i]$. The described inner free surface is convex on $[R_{gi}, r_i]$ if $z''(r) > 0$, $\forall r \in [R_{gi}, r_i]$.

Theorem 6. If there exists a solution of the eq. (8), which describes a convex inner free surface of a static meniscus on the closed interval $[R_{gi}, r_i]$ and $r_i = m \cdot R_{gi}$ with $1 < m < \dfrac{R_{gi} + R_{ge}}{2 \cdot R_{gi}}$, then the following inequalities hold:

$$-\gamma \cdot \frac{\pi/2 - (\alpha_c + \alpha_g)}{(m-1) \cdot R_{gi}} \cdot \cos\alpha_c - \frac{\gamma}{R_{gi}} \cdot \cos\alpha_g \leq p_i$$

$$\leq -\gamma \cdot \frac{\pi/2 - (\alpha_c + \alpha_g)}{(m-1) \cdot R_{gi}} \cdot \sin\alpha_g + \rho \cdot g \cdot R_{gi} \cdot (m-1) \cdot \tan\left(\pi/2 - \alpha_g\right) - \frac{\gamma}{m \cdot R_{gi}} \cdot \sin\alpha_c$$

(A.1.6)

Theorem 7. Let m be such that $1 < m < \dfrac{R_{gi} + R_{ge}}{2 \cdot R_{gi}}$. If p_i satisfies the inequality:

$$p_i < -\gamma \cdot \frac{\pi/2 - (\alpha_c + \alpha_g)}{(m-1) \cdot R_{gi}} \cdot \cos\alpha_c + \frac{\gamma}{R_{gi}} \cdot \cos\alpha_g$$

(A.1.7)

then there exists $r_i \in [R_{gi}, m \cdot R_{gi}]$, such that the solution of the initial value problem:

$$\begin{cases} z'' = \dfrac{\rho \cdot g \cdot z - p_i}{\gamma} \cdot \left[1 + (z')^2\right]^{\frac{3}{2}} - \dfrac{1}{r} \cdot \left[1 + (z')^2\right] \cdot z' \text{ for } R_{gi} < r \le \dfrac{R_{gi} + R_{ge}}{2} \\ z(R_{gi}) = 0, \ z'(R_{gi}) = \tan \alpha_c \end{cases} \quad \text{(A.1.8)}$$

on the interval $[R_{gi}, r_i]$ describes the convex inner free surface of a static meniscus.

Corollary 8. If for p_i the following inequality holds,

$$p_i < -2 \cdot \gamma \cdot \frac{\pi/2 - (\alpha_c + \alpha_g)}{R_{ge} - R_{gi}} \cdot \cos \alpha_c - \frac{\gamma}{R_{gi}} \cdot \cos \alpha_g \quad \text{(A.1.9)}$$

then there exists $r_i \in \left(R_{gi}, \dfrac{R_{gi} + R_{ge}}{2}\right)$ (close to $\dfrac{R_{gi} + R_{ge}}{2}$) such that the solution of the i.v.p. (A.1.8) on the interval $[R_{gi}, r_i]$ describes a convex inner free surface of a static meniscus.

Corollary 9. If for $1 < m' < m < \dfrac{R_{gi} + R_{ge}}{2 \cdot R_{gi}}$ the following inequalities hold

$$-\gamma \cdot \frac{\pi/2 - (\alpha_c + \alpha_g)}{(m'-1) \cdot R_{gi}} \cdot \sin \alpha_g + \rho \cdot g \cdot R_{gi} \cdot (m'-1) \cdot \tan\left(\pi/2 - \alpha_g\right) - \frac{\gamma}{m' \cdot R_{gi}} \cdot \sin \alpha_c \quad \text{(A.1.10)}$$

$$< p_i < -\gamma \cdot \frac{\pi/2 - (\alpha_c + \alpha_g)}{(m-1) \cdot R_{gi}} \cdot \cos \alpha_c - \frac{\gamma}{R_{gi}} \cdot \cos \alpha_g$$

then there exists r_i in the interval $[m' \cdot R_{gi}, m \cdot R_{gi}]$ such that the solution of the i.v.p. (A.1.8) on the interval $[R_{gi}, r_i]$ describes a convex inner free surface of a static meniscus.

Theorem 10. If a solution $z_1 = z_1(r)$ of the eq. (8) describes a convex inner free surface of a static meniscus on the interval $[R_{gi}, r_i]$, then it is a weak minimum for the energy functional of the melt column (9).

8. Appendix 2. Inequalities for single crystal tube growth by E.F.G. technique - Concave outer and inner free surface

Consider the equation (6) for $0 < R_{gi} < \dfrac{R_{gi} + R_{ge}}{2} \le r_e < R_{ge}$, α_c, α_g such that $\alpha_g \in \left(0, \pi/2\right)$, $0 < \pi/2 - \alpha_g < \alpha_c < \pi/2$.

Definition 3. The outer free surface is concave on $[r_e, R_{ge}]$ if $z''(r) < 0$, $\forall r \in [r_e, R_{ge}]$.

Theorem 11. If there exists a concave solution $z_e = z_e(r)$ of the equation (6) then $n = \dfrac{R_{ge}}{r_e}$ and p_e satisfy the following inequalities:

$$
\frac{n}{n-1} \cdot \gamma \cdot \frac{\alpha_c + \alpha_g - \pi/2}{R_{ge}} \cdot \cos \alpha_c + \frac{\gamma}{R_{ge}} \cdot \cos \alpha_g \le p_e \le \frac{n}{n-1} \cdot \gamma \cdot \frac{\alpha_c + \alpha_g - \pi/2}{R_{ge}} \cdot \sin \alpha_g
$$
$$
+ \frac{n-1}{n} \cdot \rho \cdot g \cdot \tan \alpha_c + \frac{n \cdot \gamma}{R_{ge}} \cdot \sin \alpha_c
$$

(A.2.1)

Theorem 12. If for $1 < n' < n < \dfrac{2 \cdot R_{ge}}{R_{ge} + R_{gi}}$ and p_e the inequalities hold:

$$
\frac{n}{n-1} \cdot \gamma \cdot \frac{\alpha_c + \alpha_g - \pi/2}{R_{ge}} \cdot \sin \alpha_g + \frac{n-1}{n} \cdot \rho \cdot g \cdot R_{ge} \cdot \tan \alpha_c + \frac{n \cdot \gamma}{R_{ge}} \cdot \sin \alpha_c < p_e <
$$
$$
\frac{n'}{n'-1} \cdot \gamma \cdot \frac{\alpha_c + \alpha_g - \pi/2}{R_{ge}} \cdot \cos \alpha_c + \frac{\gamma}{R_{ge}} \cdot \cos \alpha_g.
$$

(A.2.2)

then there exists $r_e \in \left[\dfrac{R_{ge}}{n}, \dfrac{R_{ge}}{n'} \right]$ and a concave solution of the equation (6).

Theorem 13. A concave solution $z_e(r)$ of the equation (6) is a weak minimum of the free energy functional of the melt column (7).

Consider now the differential equation (8) for $0 < R_{gi} < r_i < \dfrac{R_{gi} + R_{ge}}{2} < R_{ge}$ and α_c, α_g such that $0 < \pi/2 - \alpha_g < \alpha_c < \pi/2$, $\alpha_g \in \left(0, \pi/2\right)$.

Theorem 14. If there exists a concave solution $z_i = z_i(r)$ of the equation (8) then $m = \dfrac{r_i}{R_{gi}}$ and p_i satisfies the following inequalities:

$$
\frac{1}{m-1} \cdot \gamma \cdot \frac{\alpha_c + \alpha_g - \pi/2}{R_{gi}} \cdot \cos \alpha_c - \frac{\gamma}{R_{gi}} \cdot \sin \alpha_c \le p_i \le \frac{1}{m-1} \cdot \gamma \cdot \frac{\alpha_c + \alpha_g - \pi/2}{R_{gi}} \cdot \sin \alpha_g
$$
$$
+ (m-1) \cdot \rho \cdot g \cdot R_{gi} \cdot \tan \alpha_c - \frac{\gamma}{m \cdot R_{gi}} \cdot \cos \alpha_g
$$

(A.2.3)

Theorem 15. If for $1 < m' < m < \dfrac{2 \cdot R_{ge}}{R_{ge} + R_{gi}}$ and for p_i the following inequalities hold:

$$\frac{1}{m-1} \cdot \gamma \cdot \frac{\alpha_c + \alpha_g - \frac{\pi}{2}}{R_{ge}} \cdot \sin\alpha_g + (m-1)\cdot \rho \cdot g \cdot R_{gi}\tan\alpha_c -$$

$$\frac{\gamma}{m\cdot R_{gi}} \cdot \cos\alpha_g < p_i < \frac{1}{m'-1}\cdot\gamma\cdot\frac{\alpha_c+\alpha_g-\frac{\pi}{2}}{R_{gi}}\cdot\cos\alpha_c + \frac{\gamma}{R_{gi}}\cdot\sin\alpha_c \qquad (A.2.4)$$

then there exists r_i in the interval $\left[m'\cdot R_{gi}, m\cdot R_{gi}\right]$ and a concave solution of the eq. (8).

Theorem 16. A concave solution $z_i(r)$ of the equation (8) is a weak minimum of the free energy functional of the melt column (9).

9. References

St. Balint, A.M. Balint (2009), *On the creation of the stable drop-like static meniscus, appropriate for the growth of a single crystal tube with prior specified inner and outer radii*, Mathematical Problems in Engineering, vol. 2009, Article ID:348538 (2009), pp 1-22

St. Balint, A.M. Balint (2009), *Inequalities for single crystal tube growth by edge-defined film-fed (E.F.G.) technique* , Journal of Inequalities and Applications, vol.2009, Article ID: 732106, pp.1-28

St.Balint, A.M.Balint, L.Tanasie (2008) - *The effect of the pressure on the static meniscus shape in the case of tube growth by edge-defined film-fed growth (E.F.G.) method*, Journal of Crystal Growth, Vol. 310, pp.382-390

St. Balint, L.Tanasie (2008), *Nonlinear boundary value problems for second order differential equations describing concave equilibrium capillary surfaces*, Nonlinear Studies 15, Vol. 4, pp.277-296.

St.Balint, L.Tanasie(2011), *Some problems concerning the evaluation of the shape and size of the meniscus occurring in silicon tube growth* - Mathematics in Engineering, Science and Aerospace Vol. 2, pp. 53-70.

St.Balint, L.Tanasie (2010), *A procedure for the determination of the angles* $\tilde{\alpha}_e^0(r_e,h_e;p_e)$ *and* $\tilde{\alpha}_i^0(r_i,h_i;p_i)$ *which appears in the nonlinear system of differential equations describing the dynamics of the outer and inner radius of a tube, grown by the edge-defined film-fed growth (EFG) technique*, Nonlinear Analysis: Real World Applications, Vol. 11(Issue 5), pp 4043-4053

St.Balint, L.Tanasie (2011), *The choice of the pressure of the gas flow and the melt level in silicon tube growth,* Mathematics in Engineering, Science and Aerospace, Vol. 4, pp.

H.Behnken, A.Seidl and D.Franke (2005), *A 3 D dynamic stress model for the growth of hollow silicon polygons*, Journal of .Crystal Growth, Vol 275, pp. e375-e380.

A.V.Borodin, V.A.Borodin, V.V.Sidorov and I.S.Petkov (1999), *Influence of growth process parameters on weight sensor readings in the Stepanov (EFG) technique*, Journal of .Crystal Growth, Vol. 198/199, pp.215-219.

A.V.Borodin, V.A.Borodin and A.V.Zhdanov (1999), *Simulation of the pressure distribution in the melt for sapphire ribbon growth by the Stepanov (EFG) technique*, Journal of .Crystal Growth, Vol. 198/199, pp.220-226.

L.Erris, R.W.Stormont, T.Surek, A.S.Taylor (1980), *The growth of silicon tubes by the EFG process*, Journal of .Crystal Growth, Vol. 50, pp.200-211.

R. Finn, *Equilibrium capillary surfaces* (1986), Vol. 284, Grundlehren der mathematischen Wissenschaften, Springer, New York, NY, USA.

I.P.Kalejs, A.A.Menna, R.W.Stormont and I.W.Hudrinson (1990), *Stress in thin hollow silicon cylinders grown by the edge-defined film-fed growth technique*, Journal of .Crystal Growth, Vol 104, pp.14-19.

H.Kasjanow, A.Nikanorov, B.Nacke, H.Behnken, D.Franke and A.Seidl (2007), 3Dcoupled electromagnetic and thermal modeling of EFG silicon tube growth, Journal of .Crystal Growth, Vol. 303, pp.175-179.

B.Mackintosh, A.Seidl, M.Quellette, B.Bathey, D.Yates and J.Kalejs (2006) *Large silicon crystal hollow-tube growth by the edge-defined film-fed growth (EFG) method*, Journal of .Crystal Growth, Vol 287, pp.428-432.

S.Rajendram, M.Larousse, B.R. Bathey and J.P.Kalejs (1993), *Silicon carbide control in the EFG system*, Journal of .Crystal Growth, Vol. 128, pp.338-342.

S.Rajendram, K.Holmes and A.Menna (1994), *Three-dimensional magnetic induction model of an octagonal edge-defined film-fed growth system*, Journal of .Crystal Growth, Vol. 137(No 1-2), pp.77-81.

S.N.Rossolenko (2001), *Menisci masses and weights in Stepanov (EFG) technique: ribbon, rod, tube*, Journal of .Crystal Growth, vol. 231, pp.306-315.

A.Roy, B.Mackintosh, J.P.Kalejs, Q.S.Chen, H.Zhang and V.Prasad(2000), *A numerical model for inductively heated cylindrical silicon tube growth system*, Journal of Crystal Growth , Vol 211, pp.365-371.

A.Roy, H.Zhang, V.Prasad, B.Mackintosh, M.Quellette and J.Kalejs (2000), *Growth of large diameter silicon tube by EFG technique: modeling and experiment*, Journal of .Crystal Growth , Vol 230, pp.224-231.

D.Sun, Ch.Wang, H.Zhang, B.Mackintosh, D.Yates and J.Kalejs (2004), *A multi-block method and multi-grid technique for large diameter EFG silicon tube growth*, Journal of .Crystal Growth, Vol 266, pp. 167-174

T.Surek, B.Chalmers and A.I.Mlavsky (1977), *The edge film-fed growth of controlled shape crystals*, Journal of .Crystal Growth, Vol. 42, pp.453-457

J.C.Swartz, T.Surek and B.Chalmers (1975), *The EFG process applied to the growth of silicon ribbons*, J.Electron.Mater, Vol. 4, pp.255-279.

L.Tanasie, St.Balint (2010) , *Model based, pulling rate, thermal and capillary conditions setting for silicon tube growth*, Journal of Crystal Growth, vol. 312, pp. 3549-3554

V.A.Tatarchenko (1993), *Shaped crystal growth*, Kluwer Academic Publishers, Dordrecht.

B.Yang, L.L.Zheng, B.MacKintosh, D.Yates and J.Kalejs (2006), *Meniscus dynamics and melt solidification in the EFG silicon tube growth process*, Journal of .Crystal Growth, Vol. 293, pp.509-516.

6

Crystallization in Microemulsions: A Generic Route to Thermodynamic Control and the Estimation of Critical Nucleus Size

Sharon Cooper, Oliver Cook and Natasha Loines

Durham University

UK

1. Introduction

Crystallization is ubiquitous. It is evident in natural processes such as biomineralization and gem formation, and is important in industrial processes both as a purification step and in the production of materials with specific properties, including drug polymorphs, cocrystals, mesocrystals, quasicrystals, quantum dots and other inorganic nanocrystals. Consequently it is essential to gain greater understanding of the process to be able to elicit more control over its outcome.

Crystallization occurs from melts that are supercooled, i.e. cooled below their equilibrium melting temperatures, T_{eq}. For crystallization from solution, the solutions must be supersaturated, i.e. have solute concentrations above their saturation values, c_{eq}, defined as the solute concentration in equilibrium with the macroscopic crystal. The supersaturation is the driving force for crystallization, being the difference in chemical potential, $\Delta\mu$, between the parent (melt or solution) and daughter (new crystal) phases. For crystallization from the melt, $\Delta\mu = \Delta_{fus}H\Delta T / T_{eq}$, where $\Delta_{fus}H$ is the enthalpy of fusion and $\Delta T = T_{eq} - T$, is the supercooling with T denoting the temperature. Here it is assumed that $\Delta_{fus}H$ is invariant between T and T_{eq}. For an ideal solution, the supersaturation is $\Delta\mu = kT\ln\left(c / c_{eq}\right)$, where k is the Boltzmann constant, and c/c_{eq}, is the ratio of the solute concentration compared to its saturation value, which is known as the supersaturation ratio.

The formation of any new phase from a bulk parent phase requires the creation of an interface between the two phases, which requires work. Hence there exists an energy barrier to the formation of the new phase. The process of overcoming this energy barrier is known as nucleation. In crystallization, once nucleation has occurred, crystal growth onto the nuclei proceeds until the supersaturation is relieved. Owing to this nucleation stage, crystallization from the bulk melt or solution is typically under kinetic control, with metastable forms often crystallizing initially in accordance with Ostwald's rule of stages (Ostwald, 1897). In contrast, microemulsions have the unique ability to generically exert thermodynamic control over the crystallization process. This provides significant advantages; the size of the critical nucleus can be estimated with good accuracy under thermodynamic control conditions and importantly, the stable form of a material can be identified and readily produced under ambient conditions. In this chapter we discuss the scientific rationale for this thermodynamic control and provide practical examples.

2. Theoretical considerations

2.1 Classical Nucleation Theory (CNT)

Nucleation can be modelled most simply using classical nucleation theory (CNT). Gibbs thermodynamic treatment of liquid nucleation from a vapour (Gibbs, 1876, 1878) shows that the free energy change, ΔF, involved in producing a spherical liquid nucleus from the vapour is given by:

$$\Delta F = -n\Delta\mu + \gamma A = -\frac{4\pi r^3 \Delta\mu}{3v_c} + 4\pi r^2 \gamma \tag{1}$$

where n is the number of molecules in the nucleus, $\Delta\mu$ denotes the chemical potential difference between the vapour and the liquid nucleus which defines the supersaturation, γ denotes the surface tension between the nucleus and the surrounding vapour, A denotes the surface area of the nucleus, r denotes its radius and v_c denotes the molecular volume of the new condensed phase, i.e. the liquid. This same thermodynamic treatment is often used for crystallization. For crystallization from bulk solutions at constant pressure, the relevant free energy to use is the Gibbs free energy. For crystallization from microemulsions, however, there is a very small Laplace pressure difference across the microemulsion droplet interface and so the Helmholtz free energy for constant volume systems is the appropriate free energy to employ.

Equation (1) clearly shows that the favourable formation of the bulk new phase in any supersaturated system, given by the first term $-n\Delta\mu$, is offset by the unfavourable surface free energy term, γA, that necessarily arises from creating the new interface. The surface free energy term dominates at smaller nucleus sizes and leads to the nucleation energy barrier (see Figure 1a). In particular, differentiating equation (1) with respect to r leads to a maximum at

$$r^* = 2\gamma v_c / \Delta\mu, \tag{2}$$

i.e. the well-known Gibbs-Thomson equation, with the nucleation barrier given by:

$$\Delta F^* = \frac{16}{3\Delta\mu^2}\pi\gamma^3 v_c^2. \tag{3}$$

The r^* nucleus is termed the critical nucleus. It is of pivotal importance in CNT because it determines the size above which it is favourable for the new phase to grow. Nuclei smaller than r^* have a greater tendency to dissolve (or melt) than grow, whilst nuclei larger than r^* will tend to grow. The r^* nucleus has an equal probability of growing or dissolving (melting) and is in an unstable equilibrium with the surrounding solution (or melt). At larger r, a stable r_0 nucleus with $\Delta F = 0$ occurs (see Figure 1a).

Equation (3) gives the magnitude of the nucleation barrier when the new phase forms within the bulk parent phase, which is known as homogeneous nucleation. If the new phase forms on a foreign surface, however, heterogeneous nucleation occurs. The corresponding heterogeneous nucleation barrier, ΔF_{het} that arises from forming a cap-shaped critical nucleus with a contact angle, θ, on the foreign surface is given by:

$$\Delta F_{het}^{*} = \frac{16\pi\gamma^{3}v_{c}^{2}}{3\Delta\mu^{2}}f(\theta) = \Delta F_{hom}^{*}\frac{v_{het}^{*}}{v_{hom}^{*}} \tag{4}$$

where: ΔF_{hom}^{*} denotes the homogeneous nucleation barrier of the system given by equation (3), $f(\theta) = (2 - 3\cos\theta + \cos^{3}\theta)/4$ and v_{het}^{*} and v_{hom}^{*} denote the volumes of the heterogeneous and corresponding homogeneous critical nuclei, respectively. Note that equation (4) ignores the entropic free energy contribution arising from the number of surface sites upon which the nucleus may form, as this factor is incorporated in the pre-exponential factor, Ω, instead.

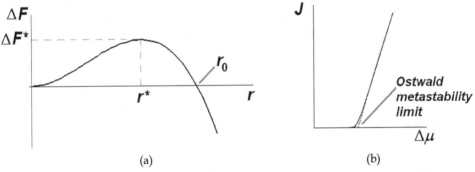

(a)   (b)

Fig. 1. (a) Schematic graph of free energy change, ΔF vs. nucleus size, r, for a homogeneously nucleating crystal showing the critical nucleus, r^{*}, and stable nucleus, r_{0}, sizes. (b) Schematic graph of nucleation rate, J, vs. supersaturation, $\Delta\mu$, showing the Ostwald metastability limit which gives the onset crystallization temperature, T_{c}.

The kinetic theory of CNT (Volmer & Weber, 1926; Becker & Döring, 1935) used the nucleation barrier, ΔF^{*}, in an Arrhenius-type equation to derive the nucleation rate, J as:

$$J = \Omega\exp(-\Delta F^{*}/kT) \tag{5}$$

where Ω is the pre-exponential factor accounting for the rate at which the molecules impinge upon, and are incorporated into, the critical nuclei. The form of equation (5) is such that J remains negligibly small until the supersaturation reaches a critical value, the Ostwald metastability limit, at which point J suddenly and dramatically increases (see Figure 1b). Hence, an onset crystallization temperature, T_{c}, can be identified with this metastability limit, and the corresponding nucleation rate can be set, with little loss in accuracy, to a suitable detection limit for the technique monitoring the crystallization.

CNT is widely adopted because of its simplicity. However this simplicity limits its ability to model real systems. Two main assumptions are: firstly that it considers the nuclei to be spherical with uniform density and a structure equivalent to the bulk phase, and secondly that the nuclei interfaces are infinitely sharp and have the same interfacial tensions as found at the corresponding planar interfaces. A recent review by Erdemir et al., 2009, details all the assumptions of CNT, and its applicability to different systems. Notably, given that CNT stems from the condensation of a liquid from its vapour, it cannot model two stage nucleation (Vekilov, 2010), where solute molecules organize initially into an amorphous cluster, from which long range crystal order then emerges on cluster rearrangement. Despite these many limitations, CNT is useful to benchmark crystallization experiments because this approach has been so widely adopted. More importantly here, it allows useful insights into the crystallization process that are readily apparent due to its simplicity. In particular, the use of CNT has enabled us to establish that thermodynamic control of crystallization is possible in 3D nanoconfined volumes, as shown below.

2.1.1 Adoption of CNT to curved interfaces

For crystallization in nanodroplets, the planar substrate of CNT's heterogeneous nucleation formulation is replaced by a highly curved concave substrate. For such curved substrates, the free energy becomes (Cooper et al., 2008; Fletcher, 1958):

$$\Delta F_{het} = -\frac{4}{3v_c}\pi r^3 \Delta\mu[(f(\theta+\phi)-(R/r)^3 f(\phi)]+2[1-\cos(\theta+\phi)]\pi r^2 \gamma - 2\cos\theta(1-\cos\phi)\pi R^2 \gamma \quad (6)$$

where θ is the contact angle between the nucleus and the spherical substrate, ϕ is the angle between the spherical substrate and the plane connecting the nucleus edge and $f(\alpha) = 0.25(2-3\cos\alpha+\cos^3\alpha)$ (see Figure 2a). Note that for concave surfaces, corresponding to crystallization within the curved substrate, R and ϕ are assigned negative values.

The maximum in ΔF_{het} gives the barrier to nucleation, ΔF_{het}^*, and again this condition is satisfied when $r^* = 2\gamma v_c / \Delta\mu$ to give:

$$\Delta F_{het}^* = \frac{\Delta F_{hom}^*}{2}\left\{1-3x^2\cos\theta+2x^3+y(1+x\cos\theta-2x^2)\right\} = \Delta F_{hom}^* f(\theta_p) \quad (7)$$

where $x = R/r^*$, $y = \pm(x^2-2x\cos\theta+1)^{0.5}$ with the positive and negative roots applying to a nucleus on a convex and concave surface, respectively and $\cos\theta_p = x-y$. θ_p is the angle between the corresponding planar critical nucleus and the plane tangential to the curved substrate surface, as shown in Figure 2a.

Equation (7) shows that at a given temperature, and hence constant ΔF_{hom}^* value, ΔF_{het}^* depends only upon the θ_p value. Consequently in Figure 2b, all the spherical substrates depicted result in the same ΔF_{het}^* value. This can be rationalized as follows. For nucleation on a concave surface, the critical nucleus volume, v^*, is reduced compared to the planar case, and hence fewer molecules need to cluster together to form the critical nucleus. However this effect is negated by the greater contact angle, θ, compared to θ_p, which means that more work is required to create unit area of the nucleus-substrate interface, and so the mean energy increase on addition of a molecule to the sub-critical nucleus is larger. In contrast on a convex substrate, v^* is increased compared to the planar case, but θ is decreased.

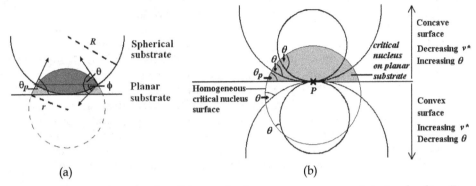

(a)   (b)

Fig. 2. (a) Schematic diagram describing nucleation upon a concave substrate of radius, R. The dark grey regions depict the nucleus on the concave surface. (b) Schematic diagram showing that for a given supersaturation, and hence critical nucleus radius, r^*, all surfaces through point P that cross the homogeneous critical nucleus surface produce the same ΔF^*_{het} value for nucleation, since they all have the same θ_p value (Cooper et al, 2008).

The onset temperature for the phase transition, e.g. the highest temperature, T_{trans}, at which crystallization should be observable, can then be found by setting the nucleation rate, J_{trans}, at T_{trans} to a suitable detection limit, where $J_{trans} = \Omega \exp(-\Delta F^*/kT_{trans})$. The pre-exponential factor, Ω, can be considered constant provided the temperature range is narrow, since ΔF^* has the greater temperature dependence. Using $(\Delta F^*/k) = (T_{eq} - \Delta T_{trans})\ln(\Omega/J_{trans})$, where $\Delta T_{trans} = T_{eq} - T_{trans}$, and substituting in equations (3), (5) and (7), gives:

$$\Delta T_{trans}^{\ 3} - \Delta T_{trans}^{\ 2}T_{eq} + \frac{16\pi\gamma^3 v_c^{\ 2}T_{eq}^{\ 2}f(\theta_p)}{3k\Delta_{fus}H^2\ln(\Omega/J_{trans})} = 0 .\qquad(8)$$

This equation has three roots corresponding to (1) The onset crystallization temperature T_c, (2) the expected onset melting temperature, T_m, for a nucleation-based melting transition, which would be required if surface melting did not occur, and (3) a non-physical root $T_c = (2T_{eq}/3)\{1-\cos W\}$ close to 0 K corresponding to the case where the critical nucleus contains only one molecule and the energy barrier is vanishingly small. T_c and T_m are given by:

$$T_{c,m} = \frac{T_{eq}}{3}\{2 + \cos W \mp (3^{0.5}\sin W)\}\qquad(9)$$

where $W = \dfrac{1}{3}\arccos\left[1 - \dfrac{72\pi\gamma^3 v_c^{\ 2}f(\theta_p)}{k\Delta_{fus}H^2 T_{eq}\ln(\Omega/J_{trans})}\right]$.

Hence T_c and T_m can be found with $x, \theta, T_{eq}, v_c, \gamma, \Delta_{fus}H$ and Ω/J_{trans} as input. r^* and R can then be obtained from the Gibbs-Thomson equation, and $R = r^* x$, respectively.

From equation (7), we find that for a constant contact angle, θ, the energy barrier to nucleation is smaller for a concave surface than a convex one, and that the reduction in energy increases as $|x|$ increases. The onset crystallization temperatures, T_c, obtained from equation (9) are therefore correspondingly higher for concave surfaces. Figure 3 shows the expected T_c as a function of $|R|$ for the case of ice crystallization on a concave surface with a contact angle, θ, between the crystal nucleus and substrate of (a) 180°, i.e. the homogeneous nucleation case and (b) 100°. The onset melting temperatures, T_m, expected for the same systems (i.e. with supplementary contact angles between the melt-nucleus and substrate of (a) 0° and (b) 80°) in the absence of surface melting are also shown in this Figure. This melting is denoted nucleation-melting. The T_m and T_c curves meet at $|R_{min}|$ and at smaller concave radii, the melting curve falls below the crystallization one, which is clearly non-physical. Hence, equation (9) cannot model crystallization in 3D nanoconfinements smaller than $|R_{min}|$. This demonstrates a fundamental limitation of the theory, and shows that a key factor necessary for crystallization has been ignored.

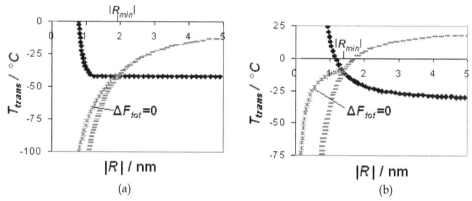

Fig. 3. The predicted ice onset crystallization temperatures, T_c, (filled diamonds) and nucleation-melting temperatures, T_m, (dashes), which would occur in the absence of surface melting, as a function of substrate radius, $|R|$, for crystallization within a spherical substrate with (a) θ = 180° and (b) θ=100°. The ice T_c have been determined using reasonable values (Pruppacher, 1995; Bartell, 1998; Speedy, 1987) of v_c = 3.26 × 10⁻²⁹ m³, γ = 20 mN m⁻¹, $\Delta_{fus}H$ = 4060 J mol⁻¹ and Ω/J_{trans} = 10¹⁸ for homogeneous nucleation and values of v_c = 3.26 × 10⁻²⁹ m³, γ = 22 mN m⁻¹, $\Delta_{fus}H$ = 5000 J mol⁻¹ and Ω/J_{trans} = 10¹⁵ for the heterogeneous nucleation case. Note that for $|R| < |R_{min}|$, the ice T_c and T_m are both given by the curve labelled ΔF_{tot} = 0, as the transition temperature is now determined by the condition that the nucleus grows to a size r_0.

2.2 Phase transitions within nanoconfined volumes (Cooper et al., 2008)

2.2.1 Crystallization from the melt

The phase transition temperature in equation (9) is determined only by the ability to surmount the nucleation barrier. The thermodynamic feasibility of the transition, i.e. whether the new phase is stable with $\Delta F \leq 0$, however, is not considered. In fact, the crossing point of the crystallization and nucleation-melting curves, occurring at a droplet size denoted R_{min}, corresponds to the system where the total phase transition of the droplet, from all liquid to all crystal, or vice versa, occurs with a free energy change $\Delta F_{tot} = 0$. For droplet sizes smaller than $|R_{min}|$, the nucleation-melting and crystallization curves describe systems where $\Delta F_{tot} > 0$, so the phase transformation would not proceed. This means that for sizes below $|R_{min}|$, the critical nucleus size can be attained as ΔF_{het}^{*} is surmountable, but there is then insufficient material within the confining substrate for ΔF to decrease to zero through further nucleus growth, e.g. in Figure 1a, the nucleus cannot grow to a size r_0. Consequently, below $|R_{min}|$ the crystallization and nucleation-melting curves must both follow the same curve labelled $\Delta F_{tot} = 0$ in the Figure to ensure a thermodynamically feasible phase transformation occurs. Hence the hysteresis normally observed upon heating and cooling the *same* system would be expected to disappear for phase transformations confined to within volumes with $|R| \leq |R_{min}|$. Unfortunately, there is a difficulty in verifying this lack of hysteresis experimentally, because this would require the r_0 nuclei to be constrained to this size. Typically, however, the r_0 nuclei subsequently grow via collisions with uncrystallized droplets, by oriented attachment of other nuclei, or by Ostwald ripening, and so it is difficult to ensure that subsequent melting and crystallization cycles are indeed performed on the same system.

Using the condition that for droplet sizes with $|R| \leq |R_{min}|$, $\Delta F_{tot} = 0$ on complete crystallization of the spherical droplet, we find:

$$R = \frac{3v_c\gamma}{\Delta\mu}\cos\theta = \frac{3}{2}r*\cos\theta \cdot \qquad (10)$$

Here we have retained the convention that the substrate radii, R, must take negative values for nucleation on a concave surface. Consequently, the critical nucleus size can be obtained from equation (10) if R and θ are known. This is an important finding because determination of $r*$ usually relies on the Gibbs-Thomson equation and the inappropriate application of bulk interfacial tension values to small nuclei. The number of molecules, $n*$, in the critical nucleus is then given by:

$$n* = \frac{v*}{v_c} = \frac{4\pi|R|^3}{3v_c}\left\{\frac{1}{2} - \frac{4}{27\cos^3\theta}\left(1 - \frac{2-(3/4)\cos^2\theta + (9/8)\cos^4\theta}{\left(4 - 3\cos^2\theta\right)^{0.5}}\right)\right\} \cdot \qquad (11)$$

The dependence of $n*$ on θ is relatively weak, so that even if θ can only be estimated to within ~10%, $n*$ is known with good precision if R can be measured. For homogeneous nucleation ($\theta = 180°$), equation (10) and (11) simplify to $r* = 2|R|/3$ and $n* = 32\pi|R|^3/81v_c$, so experimental measurement of R directly gives $r*$ and $n*$ provided $|R| \leq |R_{min}|$. So we just need to find the value of $|R_{min}|$.

An empirical determination of the R_{min} value is possible for homogeneous nucleation because the onset crystallization temperature, T_c, should be approximately constant for confinements with sizes above $|R_{min}|$. Note though that nucleation is a stochastic process, so repeated experiments will show some slight variation but an expected homogeneous nucleation temperature should nevertheless be apparent. For instance, the homogeneous nucleation temperature for ice is ≈-40 °C (Wood & Walton, 1970; Clausse et al., 1983).

Consequently, $|R_{min}|$ is readily identifiable as the droplet size at which T_c begins to decrease with decreasing $|R|$, provided the system nucleates homogeneously. For heterogeneous nucleation, θ is also likely to be a function of R, so an empirical determination is more difficult. Instead, the theoretical R_{min} value can be used, which is obtained as follows.

Substituting for r^* in equation (10) using the Gibbs-Thompson equation, $r^* = 2\gamma v_c / \Delta\mu = 2\gamma v_c T_{eq} / \Delta_{fus}H\Delta T_{trans}$, we find that for $|R| \le |R_{min}|$, T_c and T_m are given by:

$$T_c = T_m = T_{eq}\left[1 + \frac{3\gamma v_c \cos\theta}{|R|\Delta_{fus}H}\right].$$ (12)

Equation (12) has previously been identified (Couchman & Jesser, 1977) as giving the minimum possible melting temperature of a small particle. Our analysis shows that it also gives the maximum possible freezing temperature of a confined object (Vanfleet & Mochel, 1995; Enustun et al., 1990). R_{min} is then readily obtained by substituting in equation (9), since the T_c and T_m curves from equation (9) meet at R_{min}. Hence:

$$R_{min} = \frac{9\gamma v_c \cos\theta}{\Delta_{fus}H\left\{1 - \cos W + \left(3^{0.5}\sin W\right)\right\}}$$ (13)

Finding reliable R_{min} values from equation (13) requires knowledge of θ, γ and $\Delta_{fus}H$, which is problematic since the use of bulk θ, γ and $\Delta_{fus}H$ values for such small systems is likely to introduce unquantifiable errors. Consequently, the preferred methodology is that of using homogeneously nucleating systems to identify $|R_{min}|$ from the confinement size below which the onset crystallization temperatures, T_c, start to decrease. Then critical nucleus sizes can be reliably found for all sizes below $|R_{min}|$ using $r^* = 2|R|/3$ and $n^* = 32\pi|R|^3/81v_c$. Fortunately, crystallization in microemulsions often proceeds via homogeneous nucleation, making these the system of choice.

2.2.2 Crystallization from nanoconfined solution

Our crystallization model can be extended to crystallization of solutes from nanoconfined solutions, though here the situation is complicated by the decrease in supersaturation that arises as the nucleus grows. In particular, by adopting a classical homogeneous nucleation approach for crystallization from an ideal solution in a spherical confining volume, V, the free energy change, ΔF, to produce a nuclei containing n molecules would be given by (Cooper et al., 2008; Reguera et al., 2003) :

$$\Delta F = -n\Delta\mu + \gamma A + NkT\left\{\ln\left(1 - \frac{v}{Vv_c c_0}\right) - \frac{1}{v_c c_0}\ln\left(1 - \frac{v}{V}\right)\right\}$$ (14)

where $\Delta\mu = kT\ln\left(c/c_{eq}\right)$ denotes the supersaturation at that nucleus size, with $c = (N - n)/(V - v) = c_0\left\{1 - \left(v/Vv_c c_0\right)\right\}\left\{1/(1 - v/V)\right\}$, γ and A denote the interfacial tension and surface area, respectively, at the nucleus-solution interface, N is the initial number of solute molecules when $n = 0$, v denotes the nucleus volume, v_c denotes the molecular volume of the crystalline species and c_0 denotes the initial solute concentration when $n = 0$. The first two terms comprise those expected from classical nucleation theory for crystallization from unconfined volumes, whilst the third term provides the correction due to the supersaturation depletion as the nucleus grows. The free energy difference, ΔF, now exhibits a maximum, ΔF^*, corresponding to the critical nucleus radius, r^*, *and a minimum*, ΔF^*_{min}, at a larger nucleus radius, r^*_{min}, owing to the decrease in the supersaturation as the nucleus grows. r^* and r^*_{min} are both given by the usual Gibbs-Thomson equation with ΔF^* and ΔF^*_{min} both given by:

$$\Delta F^* = \frac{16\pi\gamma^3 v_c^2 T_{eq}^2 f(\theta_p)}{3\Delta_{fus}H^2 \Delta T_c^2} + NkT\left\{\ln\left(1 - \frac{v^*}{V v_c c_0}\right) - \frac{1}{v_c c_0}\ln\left(1 - \frac{v^*}{V}\right)\right\}, \tag{15}$$

where v^* denotes the nucleus volume when $r = r^*$, with the subscript *min* used to distinguish the minimum value from the maximum, and T_{eq} denotes the saturation temperature for the solution at concentration c^* surrounding the critical nucleus, r^*.

As before, the onset crystallization temperature is found from $\left(\Delta F^* / kT_c\right) = \ln(\Omega / J_{trans})$, with Ω assumed constant. This leads to a quartic equation (Cooper et al., 2008):

$$\Delta T_c^4 - T_{eq}\Delta T_c^3 + (X - Y)\Delta T_c + T_{eq}Y = 0 \tag{16}$$

where $X = \dfrac{16\pi\gamma^3 v_c^2 f(\theta_p) T_{eq}^2}{3k\ln(\Omega / J_{trans})\Delta_{fus}H^2}$

and $Y = -\dfrac{32\pi c_0}{3\ln(\Omega / J_{trans})}\left(\dfrac{\gamma v_c T_{eq} x}{\Delta_{fus}H}\right)^3\left\{\ln\left(1 - \dfrac{v^*}{V v_c c_0}\right) - \dfrac{1}{v_c c_0}\ln\left(1 - \dfrac{v^*}{V}\right)\right\}.$

Equation (16) reduces to equation (8), the melt crystallization case, when $Y = 0$. Equation (16) is solvable with x, θ, c_0, T_{eq}, v_c, γ, $\Delta_{fus}H$ and Ω / J_{trans} as input. For the typical case where $T_c \leq T_{eq}$, T_c is then given by:

$$T_c = \frac{3T_{eq}}{4} + z_1^{0.5} + z_2^{0.5} - z_3^{0.5} \tag{17}$$

where: $z_1 = \left\{-\left[\dfrac{T_{eq}(X + 3Y)}{48}\right]^{0.5}\left(\cos\left(\dfrac{Z}{3}\right) - 3^{0.5}\sin\left(\dfrac{Z}{3}\right)\right)\right\} + \dfrac{T_{eq}^2}{16}$,

$z_2 = \left\{-\left[\dfrac{T_{eq}(X + 3Y)}{48}\right]^{0.5}\left(\cos\left(\dfrac{Z}{3}\right) + 3^{0.5}\sin\left(\dfrac{Z}{3}\right)\right)\right\} + \dfrac{T_{eq}^2}{16}$,

and

$z_3 = \left\{\left[\dfrac{T_{eq}(X + 3Y)}{12}\right]^{0.5}\cos\left(\dfrac{Z}{3}\right)\right\} + \dfrac{T_{eq}^2}{16}$, where $Z = \arccos\left\{ \dfrac{27^{0.5}\left[T_{eq}{}^3 Y + (X - Y)^2 \right]}{2\left[T_{eq}(X + 3Y) \right]^{1.5}} \right\}$.

r^* and R are then found from the Gibbs-Thomson equation, and $R = r^* x$, respectively.

Of the other three solutions to the quartic equation (16), two are non-physical, as they correspond to either crystallization close to 0 K, or the onset crystallization temperature close to T_{eq} found when the minimum free energy radius, r_{min}^* , is used instead of the maximum value, r^*. The remaining solution provides the onset crystallization temperature for rare cases when $T_c > T_{eq}$, which could in principle arise for sufficiently soluble species when $\theta < 90°$. In this case, positive values of x are used since r is negative as well as R and

$$T_c = \frac{3T_{eq}}{4} + z_1^{0.5} - z_2^{0.5} + z_3^{0.5} . \tag{18}$$

As with the melt crystallization case, equations (17) and (18) are valid until the confinement size decreases to such an extent that there is insufficient crystallizing material present to ensure an energetically feasible phase transformation. For instance, crystallization would not be possible in a 3D-nanoconfined solution having the ΔF vs r curve shown in Figure 4a, since $\Delta F_{min}^* > 0$. For these small nanoconfinements, crystallization becomes feasible when the *minimum energy* ΔF_{min}^* is set to zero so that $r_{min}^* = r_0$, i.e. a stable nucleus can form. We then obtain (Cooper et al., 2008):

$$R = \frac{\gamma f(\theta_{p,min})}{kT_c c_0 x_{min}{}^2\left\{ \ln\left(1 - \dfrac{v_{min}^*}{V v_c c_0}\right) - \dfrac{1}{v_c c_0}\ln\left(1 - \dfrac{v_{min}^*}{V}\right) \right\}} = \frac{2\gamma v_c x_{min}}{kT_c \ln(c_{min}^* / c_{eq})} \tag{19}$$

from which:

$$x_{min} = \frac{R}{r_{min}^*} = \left(\frac{\ln(c_{min}^* / c_{eq}) f(\theta_{p,min})}{2 v_c c_0 \ln\left(1 - \dfrac{v_{min}^*}{V v_c c_0}\right) - 2\ln\left(1 - \dfrac{v_{min}^*}{V}\right)} \right)^{\frac{1}{3}} \tag{20}$$

where v_{min}^* denotes the volume of the r_{min}^* nucleus with $\Delta F = 0$, $\partial \Delta F / \partial r = 0$, and $\partial^2 \Delta F / \partial r^2 > 0$, and c_{min}^* denotes the solute concentration surrounding the r_{min}^* nucleus. Equation (20) can be solved iteratively to give $x_{min} = R / r_{min}^*$ with inputted values for c_0, θ, v_c and c_{eq}, but again, crucially not the γ or $\Delta_{fus}H$ values. The x value is then given by:

$$\frac{x}{\ln(c^* / c_{eq})} = \frac{x_{min}}{\ln(c_{min}^* / c_{eq})} . \tag{21}$$

Thus we can work out $r^* = R/x$ just by measuring R and T_c values, when $|R| \leq |R_{min}|$. The number of molecules, n^* $(= v^*/v_c)$ in the critical nucleus is also obtained from R, θ, c_0, v_c, and T_c as input and is given by:

$$n^* = \frac{4\pi R^3}{3v_c x^3}\left\{\frac{1}{2} - \frac{x^3}{2} + \frac{2 - 2x\cos\theta + x^2 - x^2\cos^2\theta - 2x^3\cos\theta + 2x^4}{4y}\right\}, \tag{22}$$

which for homogeneous nucleation ($\theta = 180°$, $y = x+1$) reduces to the expected $n^* = 4\pi r^{*3}/3v_c$.

Hence provided $|R| \leq |R_{min}|$, we can again determine both n^* and r^* without reliance on macroscopic γ and $\Delta_{fus}H$ values, provided we know or can estimate θ. The experimental onset crystallization temperature, T_c, can then be compared with the values predicted from the Gibbs-Thomson and ideal solubility equations using the experimentally found R, x, and c_{eq} values. This provides a measure of how bulk values of γ and $\Delta_{fus}H$ are likely to be perturbed for solute crystallization in nanosystems. Figure 4a shows a schematic graph of ΔF vs r, whist Figure 4b shows theoretical calculations using equations (17) and (19) to model the homogeneous nucleation of octadecane from dodecane solutions, illustrating again the decrease in T_c that occurs below $|R_{min}|$, from which the critical nucleus size can be estimated.

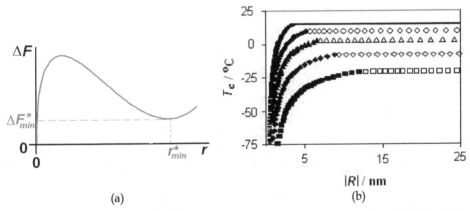

(a)   (b)

Fig. 4. (a) Schematic graph of ΔF vs r for crystallization in a 3D-nanoconfined solution. (b) Graph showing T_c, with confinement radius, $|R|$, for homogeneous nucleation from solutions of octadecane in dodecane. Open symbols correspond to the regime where $\Delta F^* = kT_c \ln(\Omega / J_{trans})$ gives T_c, filled symbols to the regime where $|R| \leq |R_{min}|$ and T_c is controlled by $\Delta F^*_{min} = 0$. Squares = 0.1 mole fraction of octadecane in dodecane, diamonds = 0.25 mole fraction, triangles = 0.5 mole fraction and circles = 0.75 mole fraction. The uppermost curve corresponds to the pure octadecane liquid case, with the thicker line portion showing the regime controlled by $\Delta F_{tot} = 0$ (Cooper et al., 2008).

2.2.3 Thermodynamic control of crystallization

The above analysis show that for all confinement sizes below $|R_{min}|$, crystallization is not governed by the ability to surmount the nucleation barrier, ΔF^*, but by the ability to create a thermodynamically feasible new phase , i.e. $\Delta_{tot}F \leq 0$ for melt crystallization or $\Delta F^*_{min} \leq 0$ for solution crystallization. This means that crystallization is under thermodynamic, rather than the usual kinetic, control. This is significant because crystallization can then be directed to generically produce the most stable crystalline phase in 3D nanoconfined solutions and liquids. This finding is particularly important for polymorphic compounds.

3. Polymorphism

Polymorphism occurs when a substance can crystallize into more than one crystal structure. Each polymorph of a substance will have differing physical properties e.g. melting points, solubilities, compaction behaviour etc. In the pharmaceutical industry, it is imperative that a drug does not transform post-marketing, as this can affect its bioavailability, and reduce the drug's effectiveness. An infamous example of this occurred for the anti-HIV drug, Ritonavir (Chemburkar et al., 2000). In 1998, 2 years after marketing the drug in the form of soft gelatine capsules and oral solutions, the drug failed dissolution tests; a less soluble, thermodynamically more stable, polymorph had formed. This resulted in the precipitation of the drug and a marked decrease in the dissolution rate of the marketed formulations, reducing its bioavailability. Consequently Ritonavir had to be withdrawn from the market and reformulated, to the cost of several hundred million dollars.

The Ritonavir case highlights that crystallization is typically under kinetic control, with metastable polymorphs often crystallizing initially in accordance with Ostwald's rule of stages (Ostwald, 1897). For pharmaceutical companies, Ostwald's rule is a nemesis, as it means that their strategy of relying on high throughput screening of different crystallization conditions in order to identify stable polymorphs is scientifically flawed and so may not succeed. Consequently the industry remains vulnerable to another Ritonavir-type crisis. If the crystallization can be conducted from 3D-nanoconfined solutions, however, the crystallization can be exerted under thermodynamic control so that the thermodynamically stable polymorph is crystallized directly. In particular, in Figure 5a it is evident that neither polymorph A in red or polymorph B in blue will crystallize from the nanoconfined solution,

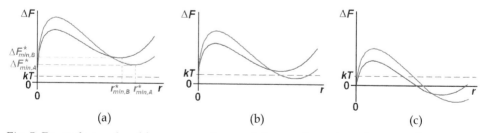

Fig. 5. Example graphs of free energy change, ΔF vs. nucleus size, r for crystallization of a polymorphic system from 3D nanoconfined solutions of monodisperse size and supersaturation. (a) System stabilized due to 3D nanoconfinement, with no observable crystallization. (b) Crystallization is under thermodynamic control with stable polymorph A (red) crystallizing, even though it has the higher nucleation barrier. (c) Crystallization occurs under kinetic control with metastable polymorph B (blue) as the majority product.

since for both $\Delta F_{\min}^{*} > kT$. In Figure 5b, however, polymorph A can form because it can produce a stable nucleus ($\Delta F_{min,A}^{*} < 0$) whereas polymorph B cannot. Hence provided the nucleation barrier is surmountable, this system will crystallize under thermodynamic control to directly give the stable polymorph A. In the system shown in Figure 5c, however, both polymorph A and B can form stable nuclei; crystallization will tend to be under kinetic control with polymorph B forming at a faster rate due to its lower nucleation energy barrier. Thus metastable B becomes the majority product in this case.

The thermodynamic arguments stated above are valid in the thermodynamic limit, where the system is so large that fluctuations do not significantly contribute to the equilibrium properties of the system. Though, of course, it must be remembered that it is these very fluctuations that enable critical nuclei to form and hence initiate the phase transformation. Consequently, in a system comprising a limited number of nanoconfined solutions, the fluctuations do need to be included to accurately model the equilibrium properties of the system (Reguera et al., 2003). We neglect this statistical thermodynamic description in our simple model of onset crystallization temperatures because the system in which we apply it, namely microemulsions, typically consists of a sufficiently large number of droplets, ~10^{18} per gram of microemulsion, which makes its additional complexity unwarranted. Moreover our simple model contains the essential features required to show how reliable estimates of critical nucleus sizes and thermodynamic control of crystallization are realizable in microemulsions.

4. Microemulsions

Microemulsions are thermodynamically stable, transparent mixtures of immiscible liquids. Typically they comprise oil droplets in water (an oil-in-water microemulsion) or water droplets in oil (a water-in-oil microemulsion). Bicontinuous microemulsions are also possible, however the absence of a 3D nanoconfined solution in these systems make them less suitable for thermodynamic control of crystallization. In the droplet microemulsions, the droplet size is typically 2-10 nm, with a relatively narrow polydispersity of $\sigma_R/R_{max} \approx 0.1$-$0.2$, where σ_R is the Gaussian distribution standard deviation and R_{max} is the modal droplet radius (Eriksson and Ljunggren, 1995). The droplets are stabilized by surfactants, frequently in combination with a co-surfactant, which reside at the droplet interface, reducing the interfacial tension there to ~10^{-3} mN m^{-1}. This ultralow interfacial tension provides the thermodynamic stability of microemulsions, since the small free energy increase involved in creating the droplet interface is more than offset by the increased entropy of the dispersed phase. Note that this ultralow interfacial tension also ensures that the LaPlace pressure difference, $\Delta P = 2\gamma/R$, across the highly curved droplets is small. When the volume fraction of the dispersed phase becomes so low that its properties differ measurably from its usual bulk properties, the terms "swollen micelles", "swollen micellar solutions", "solubilized micellar solutions" or even simply "micellar solutions" can be used instead of microemulsions for oil-in-water systems, whilst for water-in-oil systems, the same terms but with "inverse" or "reverse" inserted before "micelle" or "micellar" may be used. However, because there is, in general, no sharp transition from a microemulsion containing an isotropic core of dispersed phase and a micelle progressively swollen with the dispersed phase, many researchers use the term "microemulsion" to include swollen micelles (or swollen inverse micelles) but not micelles containing no dispersed phase. This is the context in which the term "microemulsion" is used here. In the microemulsions, dissolved solutes may be supersaturated within the discontinuous (dispersed) phase, or the dispersed liquid may be supercooled, so that crystallization in the microemulsions can occur. The solute molecules are assumed to be distributed amongst the microemulsion droplets with a Poisson distribution, so that most droplets will have a supersaturation close to the mean, but a minority will have supersaturations significantly higher than the mean.

Microemulsions are dynamic systems with frequent collisions occurring between the droplets. The most energetic of these collisions cause transient dimers to form, allowing the exchange of interior content between the droplets. This exchange means that microemulsions can act as nanoreactors for creating quantum dots and other inorganic nanoparticles, for example. A recent review (Ganguli et al., 2010) on inorganic nanoparticle formation in microemulsions highlights the progress that has been made in this area from its inception with metal nanoparticle synthesis in 1982 by Boutonnet et al., followed by its use in synthesizing quantum dots (Petit et al. 1990) and metal oxides (see e.g. Zarur & Ying, 2000). The transient dimer mechanism also enables crystallization to proceed in microemulsions whenever a transient dimer forms between a droplet containing a crystal nucleus and a nucleus-free droplet which contains supersaturated solution, since the crystal nucleus can then gain access to this supersaturated solution and thereby grow (see Figure 6). Crystallization of organic compounds from microemulsions was first studied by Füredi-Milhofer et al. in 1999 for the aspartame crystal system, with studies on glycine crystallization (Allen et al., 2002; Yano et al., 2000; Nicholson et al., 2011; Chen et al., 2011) and carbamazepine (Kogan et al., 2007) following. The use of microemulsions in producing both inorganic nanoparticles and macroscopic organic crystals shows that the size of the particulates grown can vary from a few nm to mm, depending upon the nucleation rate, the ability to form stable nuclei, and the extent of surfactant adsorption on the resulting particles. Our interest in microemulsions stems from their intrinsic ability to enable reliable estimates of critical nucleus sizes and to exert generic thermodynamic control over the crystallization process for the first time.

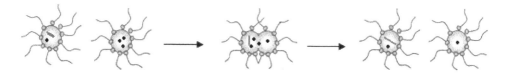

Fig. 6. Schematic diagram illustrating an energetic collision between a microemulsion droplet containing a crystal nucleus and a nucleus-free droplet containing supersaturated solution. The energetic collision results in a transient dimer forming, enabling the nucleus to gain access to more molecules and grow. The crystal nucleus is shown in red and the solute molecules are shown in black. The surfactant molecules stabilizing the microemulsion droplets are depicted as blue circles with tails.

4.1 Measurement of the critical nucleus size in microemulsions

Reliable estimates for the critical nucleus size can be found for homogeneous nucleation in microemulsions. The homogeneous nucleation temperature should be approximately constant until the droplet size decreases to below $|R_{min}|$, and thereafter the temperature should decrease. In the region where T_c decreases, the crystallization is controlled by the requirement that $\Delta_{tot}F \leq 0$, rather than the size of the nucleation energy barrier. Consequently, the critical nucleus size can be estimated by measuring the droplet size $|R|$ and assuming a spherical nucleus so that $r^* = 2|R|/3$ and $n^* = 32\pi|R|^3/81v_c$. Whilst homogeneous nucleation is comparatively rare in bulk systems, in microemulsions it is more prevalent for two main reasons. Firstly, the droplets are too small to contain foreign material onto whose surfaces heterogeneous nucleation could arise. Secondly, the ability of the surfactants to induce heterogeneous nucleation is often reduced in microemulsions compared to that at planar interfaces and in emulsions, particularly for crystallization from solution. This is because nuclei formation on the surfactant layer is disfavoured at this ultra low interfacial tension interface, and the high curvature may also hinder any templating mechanism that operates at more planar interfaces. The reduction in adsorption is readily apparent from Young's equation, $\cos\theta = (\gamma_1 - \gamma_2)/\gamma$ where θ is the contact angle, γ is the interfacial tension between the crystallizing species and the surrounding melt/solution, with γ_1 and γ_2 denoting the interfacial tensions between the surfactant and immiscible phases, and the surfactant and crystallizing species, respectively. The lowering of γ_1 that occurs in going from an emulsion to a microemulsion results in a higher contact angle, θ, and hence reduced adsorption for the crystallizing species.

Given this, we might expect heterogeneous nucleation to be impeded in microemulsions, and indeed other systems with low interfacial tensions, γ_1. Such an effect was observed at the phase inversion of an emulsion using Span 80 and Tween 80 surfactants to induce β-glycine crystallization at the oil-aqueous interface (Nicholson et al., 2005). Similarly, the ability of the nonionic surfactants, Span 80 and Brij 30 to heterogeneously nucleate glycine was negligible in microemulsions containing these mixed surfactants (Chen et al., 2011), whereas they promoted the metastable β-glycine nucleation at planar interfaces and in emulsions (Allen et al., 2002; Nicholson et al., 2005). There are literature examples where nonionic surfactants do induce heterogeneous nucleation in microemulsions, though. For instance, ice nucleation was promoted by adding heptacosanol, a long chain alcohol cosurfactant, to AOT microemulsions (Liu et al., 2007). Long chain alcohols can induce ice nucleation at temperatures of ≈-2 °C at planar air-water interfaces (Popovitz-Biro et al., 1994) and at ≈-8 °C at emulsion interfaces (Jamieson et al., 2005). This nucleating ability was diminished in the microemulsions. Nevertheless, ice crystallization still tended to occur at temperatures in the range of ≈-9 to -30 °C depending upon the heptacosanol concentration in the microemulsion droplets (Liu et al., 2007), i.e. much higher than the homogeneous nucleation temperature of ≈-40 °C (Wood & Walton, 1970; Clausse et al., 1983). For ionic surfactants, like AOT, heterogeneous nucleation in microemulsions can also occur. The longer range electrostatic interactions of ionic surfactants can induce order without requiring direct adsorption onto the surfactant layer. It is important, therefore, to choose surfactants that do not promote crystallization when placed at the planar air-liquid or air-solution interface to ensure that homogeneous nucleation occurs in the microemulsions.

Once a suitable homogeneous nucleating microemulsion system has been found, the onset crystallization temperature, T_c, for microemulsions of varying size, R, can be found by a suitable technique, such as differential scanning calorimetry (DSC). It can be assumed that the exothermic DSC peak arising from the crystallization of the droplets corresponds to the temperature range in which the majority of droplets can crystallize, so that the mean droplet size can be used to accurately determine r^* and n^*. The mean droplet size of the microemulsion can be determined from small angle X-ray scattering, or small angle neutron scattering measurements. This methodology allows a simple and accurate measurement of the critical nucleus size, and is particularly useful for crystallization of liquids, or solutes which have a high solubility in the confined phase, so that there is sufficient crystallizable material present within the microemulsion for the exothermic crystallization peak to be observable by DSC. We have recently applied this methodology to ice crystallization in AOT microemulsions (Liu et al., 2007). Figure 7a shows homogeneous ice nucleation in AOT microemulsions, compared to heterogeneous nucleation in Figure 7b where the cosurfactant heptacosanol is added to the AOT microemulsions. The larger error bars in the heterogeneous nucleation case shown in Figure 7b reflect the varying number of heptacosanol molecules in the droplets that cause ice nucleation. For the homogeneous case, the reduction in T_c with $|R|$ occurs at $|R_{min}| \approx 2$ nm in Figure 7a, in good agreement with the theoretical value shown in Figure 3a. From this, the critical nucleus at R_{min} can be estimated to contain ≈ 280 molecules (Liu et al., 2007).

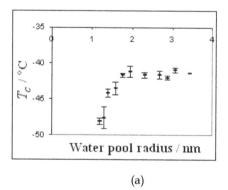

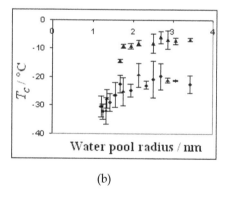

(a)   (b)

Fig. 7. Variation of observed ice onset crystallization temperatures, T_c, with water pool size for microemulsions with (a) AOT and (b) AOT plus heptacosanol The error bars show the standard deviation from three or more measurements. The nucleation is homogeneous in (a) and heterogeneous in (b). In the AOT microemulsions with added heptacosanol, several crystallization peaks were often evident, due to the droplets having differing numbers of the heptacosanol cosurfactant molecules that help nucleate ice. Consequently in (b) the highest T_c peak (upper curve) and largest T_c peak (lower curve) are just plotted for clarity (Liu et al. 2007).

4.2 Thermodynamic control of crystallization in microemulsions: Leapfrogging Ostwald's rule of stages

As detailed previously, crystallization within 3D nanoconfined solutions differs fundamentally from bulk crystallization because the limited amount of material within a nanoconfined solution results in the supersaturation decreasing substantially as the nucleus grows, leading to a minimum[11,12] in the free energy, ΔF^*_{min}, at a post-critical nanometre nucleus size, r^*_{min} (see Figure 4a). This fact, in particular, allows stable polymorphs to be solution-crystallized from microemulsions even when they have high nucleation barriers. Hence thermodynamic control of crystallization can be generically achieved for the first time. Note that microemulsions differ in two main ways from the theoretical nanoconfined solutions considered previously in sections 2 and 3. Firstly, transient droplet dimer formation enables the nuclei to grow beyond that dictated by the original droplet size; in fact crystals ranging from nm to mm can be produced. Secondly, microemulsions are polydisperse. There will be a range of droplet sizes and supersaturations in any microemulsion system, which must be considered to enable thermodynamic control of crystallization. An effective strategy is detailed below.

The equilibrium population of the r^*_{min} nuclei in the microemulsion depends upon the Boltzmann factor, $\exp\left(-\Delta F^*_{min} / kT\right)$. Consequently, if $\Delta F^*_{min} > kT$, the equilibrium population of the r^*_{min} nuclei will be very low. In contrast, if the r^*_{min} nuclei have free energies, $\Delta F^*_{min} < kT$, they will have a sizeable equilibrium population. We term such r^*_{min} nuclei, (near) stable nuclei. Crystallization in microemulsions proceeds initially via the 3D nanoconfined nuclei gaining access to more material and growing during the energetic droplet collisions that allow transient dimers to form. *This crystallization process will be severely hindered if the population of such r^*_{min} nuclei is so low* as for the case depicted in Figure 5a, that the colliding droplets are highly unlikely to contain nuclei. In this case, the supersaturated system is stabilized due to the 3D nanoconfinement. In contrast, the crystallization can proceed readily via this transient droplet dimer mechanism if $\Delta F^*_{min} < kT$

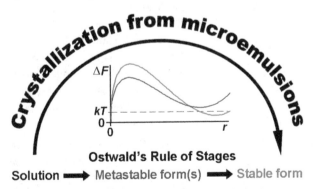

Ostwald's Rule of Stages

Solution ➡ Metastable form(s) ➡ Stable form

Fig. 8. A schematic diagram showing how Ostwald's rule of stages can be 'leapfrogged' by crystallizing from microemulsions. Stable polymorph A is in red, and metastable B is in blue. The ΔF vs r plot corresponds to the case where crystallization is brought under thermodynamic control so that stable polymorph A crystallizes directly.

because then a sizeable population of droplets will contain a (near) stable nucleus. *Thus, crystallization in microemulsions is governed by the ability to form (near) stable nuclei rather than critical nuclei.* In particular, recalling that there will be a range of droplet sizes and solute concentrations within the microemulsion droplets, thermodynamic control of crystallization can be achieved using the following methodology. The supersaturation in a microemulsion can be increased from one that is stabilized due to 3D nanoconfinement until the following condition is met: only the largest and highest supersaturation droplets, which contain the most material, can form (near) stable nuclei of only the most stable crystal form or polymorph. This situation is exemplified in Figure 8 with $\Delta F^*_{min,A} \leq kT$ but $\Delta F^*_{min,B} > kT$. Crystallization is then only just possible, and importantly, only the most stable polymorph will crystallize, as this is the only form for which (near) stable nuclei are likely to exist and so grow during transient droplet dimer formation. This generic strategy allows us to 'leapfrog' Ostwald's rule of stages and crystallize stable polymorphs directly.

The strategy detailed above will work provided $\Delta F^*_{min,A} < \Delta F^*_{min,B}$, and (near) equilibrium populations of the r^*_{min} nuclei are obtained. $\Delta F^*_{min,A} < \Delta F^*_{min,B}$ will typically be true because the stable polymorph has the greater bulk stability and it is the least soluble. Hence $r^*_{min,A} > r^*_{min,B}$ as stable polymorph A can grow to larger nucleus sizes, with typically lower free energies, than metastable B before its supersaturation is depleted. Equilibrium populations of the r^*_{min} nuclei will arise provided the r^*_{min} nuclei formation and dissolution processes are sufficiently rapid. This depends upon the magnitude of the nucleation barriers, ΔF^*, and the dissolution energy barriers, $\Delta F^* - \Delta F^*_{min}$, respectively. It can be ensured that the nucleation barriers to all polymorphs are surmountable by crystallizing from sufficiently small droplets. This is because the substantial supersaturation depletion that arises in a small droplet as the nucleus grows means that very high initial supersaturations with respect to the most stable polymorph are required to enable a (near) stable nuclei to form. Consequently it can reliably be assumed that the initial solutions in these droplets will also be sufficiently supersaturated with respect to all polymorphs that all nucleation barriers are indeed surmountable. The nuclei dissolution barriers, i.e. $\Delta F^* - \Delta F^*_{min}$, will be surmountable when the ΔF^* barriers are surmounted and $\Delta F^*_{min} \geq 0$ as in Figure 8, since then $\Delta F^* - \Delta F^*_{min} \leq \Delta F^*$. Hence thermodynamic control will indeed be obtainable if the largest and highest supersaturation droplets have free energy curves corresponding to Figure 8. Ostwald's rule will have been 'leapfrogged' because the high initial supersaturations ensure the nucleation barriers are 'leapt over' to directly give the most stable polymorph. In contrast, the analogous unconfined bulk system would crystallize the metastable polymorph initially, in accordance with Ostwald's rule, owing to its lower nucleation barrier.

Note that the ability of microemulsions to exert thermodynamic control over crystallization is independent of the nucleation path; it depends solely upon the ability to form (near) stable nuclei, rather than their formation pathway. Consequently this ability is generic, applying equally to systems that nucleate via a classical one stage mechanism, and to those where two stages are implicated.

Once formed, the (near) stable nuclei can grow via transient droplet dimer formation until the nuclei become larger than the droplets or the supersaturation is relieved. For crystallites larger than the droplets, subsequent growth can then occur via the following processes; energetic collisions with droplets that allow access to the droplets' interior contents, oriented attachment of other nuclei, and impingement from the (typically minuscule) concentration of their molecules in the continuous phase. Thus, the final crystal size can vary from nm to mm, depending upon the concentration of (near) stable nuclei, the supersaturation and the extent to which surfactant adsorption on the crystallites limits their growth rate.

In order to test the ability of microemulsions to exert thermodynamic control over crystallization, we chose three problem systems that had well-documented difficulty in obtaining their stable polymorphs: namely glycine, mefenamic acid and ROY (Nicholson et al., 2011). In each case, it was successfully demonstrated that the stable polymorph crystallized directly from the microemulsions under conditions where crystallization was only just possible. The time-scale at which ~mm sized crystals grew ranged from minutes to months. Solvent-mediated transformations to more stable polymorphs can occur in this timeframe, although such transformations would be expected to have a significantly reduced rate in microemulsions owing to the exceedingly low concentration of the polymorph's molecules in the continuous phase. Accordingly, transmission electron microscopy (TEM) was used to show that the nanocrystals crystallized within the first 24 hours in the microemulsions were also of the stable form, thereby proving that the initial crystallization was indeed in this form. TEM was performed on the microemulsions by dropping small aliquots of the microemulsions onto TEM grids, allowing the drops to (mostly) evaporate, and then washing the grids with the microemulsion continuous phase to remove residual surfactant. This left predominantly just the crystallites grown in the microemulsion droplets on the TEM grids. Figure 9 shows TEM images of stable Form I nanocrystals of mefenamic acid grown within 1 day from DMF microemulsions at 8 °C containing 80 mg/ml of mefenamic acid in the DMF (Nicholson & Cooper, 2011).

With increasing supersaturation, metastable polymorphs also crystallized from the microemulsions. The relative supersaturation increase that led to the emergence of metastable forms was highly system dependant, though. For instance, for glycine the stable γ-polymorph crystallized as the majority polymorph under mean initial supersaturation ratios of 2.0 to 2.3, for mefenamic acid the corresponding range was 4.1 to 5.3, whereas for ROY a much larger range of $\approx$10 to 24 was found. The small supersaturation range in which the stable γ-glycine polymorph crystallized as the majority form reflected the small relative energy difference of 0.2 kJ mol between the γ- and α-polymorphs and the much faster growth rate (~500 times) of the α-polymorph in aqueous solutions (Chew et al., 2007). Recall that when crystallization is only just possible in the microemulsions, the formation of (near) stable nuclei will be confined to only the largest droplets with the highest supersaturations. Hence the actual initial supersaturations that are required for crystallization to be just possible are always much higher than the mean initial values. An estimate of this actual initial supersaturation was found in the glycine system as follows. Assuming a Poisson distribution of solute molecules amongst the droplets, then for the glycine system, <10^{-8} of the droplets formed (near) stable nuclei under conditions where crystallization was only just possible, since the 0.2 kJ mol^{-1} stability difference between the γ- and α-forms would only lead to thermodynamic preference for the stable γ-form if the (near) stable nuclei contained ~20-30 molecules (Nicholson et al., 2011). This meant initial supersaturation ratios of ~11-15 were necessary for crystallization to be possible in the glycine system. A similar analysis for the mefenamic acid case gives initial supersaturation ratios >15 (Nicholson & Cooper, 2011).

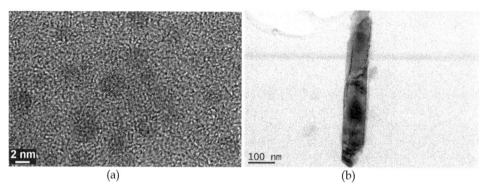

(a)   (b)

Fig. 9. TEM images of stable Form I nanocrystals of mefenamic acid grown from DMF microemulsions. (a) ~4 nm nanocrystals grown in 12 hours and (b) a Form I nanocrystal grown in 24 hours.

These very high initial supersaturation ratios highlight two key factors governing the solution crystallization of stable polymorphs from microemulsions. Firstly, the substantial supersaturation decrease as a nucleus grows in a droplet means that very high initial supersaturations are needed for a (near) stable nucleus to form. Secondly, these very high initial supersaturations help ensure that the nucleation barriers to all possible polymorphs are surmountable. Hence, the use of microemulsions is the only way to crystallize a stable polymorph that has a very high nucleation barrier in bulk solution.

4.3 Practical considerations for choosing microemulsion systems

In choosing suitable microemulsion systems for achieving thermodynamic control of crystallization to obtain stable polymorphs, or for obtaining reliable estimates of critical nucleus sizes, the following should be considered.

1. The material to be crystallized should be insoluble, or only sparingly soluble in the microemulsion continuous phase. Since the material will have the same chemical potential at equilibrium in the continuous phase, as in the confined phase, the supersaturated material could potentially crystallize in the continuous phase with attendant loss of thermodynamic control. If the crystallizing material has a very low/negligible solubility in the continuous phase, however, the impingement rate onto a nucleus in this phase is so low that the nucleation rate and subsequent growth of such nuclei will be minimized. Then it can be reliably assumed that the crystallization is initially confined to the dispersed phase so that thermodynamic control is possible.

2. The nucleation should ideally be homogeneous. Choosing surfactants that do not promote crystallization at planar or emulsion interfaces helps ensure this. Homogeneous nucleation enables critical nucleus sizes to be obtained more reliably, since the contact angle of $180°$ is, of course, invariant with droplet size, R, and the relationships $r^* = 2|R|/3$ and $n^* = 32\pi|R|^3/81v_c$ are valid for droplet sizes

$|R| \le |R_{min}|$, i.e. where T_c decreases with $|R|$. Homogeneous nucleation is also preferred when aiming to crystallize stable polymorphs, since heterogeneous nucleation could potentially lead to a metastable polymorph having a lower ΔF^*_{min} than the stable one, and thereby crystallizing in preference to the stable form.

3. Given that a metastable polymorph can potentially have the lowest ΔF^*_{min} in a microemulsion if, for example, it is heterogeneously nucleated by the surfactant, or it is sufficiently stabilized by the surrounding solvent, or an inversion of stability between polymorphs occurs at nm sizes, then this possibility should be checked. This can be done readily by using a different solvent and/or surfactant. In addition, the supersaturation of the microemulsion should be gradually increased from the point at which crystallization is only just possible, until crystals/nanocrystals of more than one polymorph crystallize. In this way, all low energy polymorphs can be identified.

4. The crystallizable species, or more often the solvent, may be absorbed in the surfactant interfacial layer and so the solute concentration within the interior pool of the microemulsion droplet may differ substantially from the bulk concentration used in making the microemulsion. This possibility must be checked by measuring the solubility of the crystallizing species in the microemulsion, and then determining the mean initial supersaturation ratios accordingly. Accurate solubility measurements require adding powdered material to a microemulsion and leaving for a prolonged period (weeks) to ensure equilibration.

5. Bicontinuous and percolating microemulsions are not suitable systems for determining critical nucleus sizes or obtaining thermodynamic control of crystallization. In bicontinuous microemulsions, the absence of 3D nanoconfinement precludes their use. In percolating microemulsions, the droplets cluster and transiently form much larger droplets in which (near) stable nuclei of metastable polymorphs can form and grow, alongside the (near) stable nuclei of the stable form in the non-clustering droplets. For water-in-oil microemulsions, the absence of percolation and bicontinuous structures can be assumed if the microemulsions show minimal conductivity.

6. Microemulsions are thermodynamically stable and so form spontaneously upon mixing the constituents. Hence shaking by hand and vortexing are suitable preparation methods. Prolonged sonication should be avoided in case this affects the crystallization.

7. When the supersaturation in the microemulsions is achieved via anti-solvent addition or by a chemical reaction, a mixed microemulsion method should generally be implemented whereby two microemulsions are prepared. A different reactant is in each microemulsion, or for the antisolvent crystallization method, one microemulsion contains the undersaturated crystallizable species in a good solvent and the other microemulsion contains the antisolvent. On mixing the two microemulsions, transient dimer formation between droplets containing different reactants and/or solvents enables a relatively rapid equilibration of interior droplet content to take place, on the timescale of ~µs to ms (Ganguli et al., 2010). After this, it can be assumed that the dispersed reactants and solvents are distributed amongst the droplets with a Poisson distribution. Adding the antisolvent or second reactant drop-wise to the microemulsion should be avoided as this can create high concentration fluctuations immediately after the addition (Chen et al., 2011). Alternatively, if one of the reactants is soluble in the continuous phase, a solution of this reactant should be added to the microemulsion

containing the second reactant. Here, it is necessary to ensure the reaction proceeds predominantly in the microemulsion droplet, or at the droplet interface, by making sure that the second reactant has a negligible solubility in the continuous phase, whilst the first reactant must be able to partition into the droplet interface and/or interior.

8. For determining critical nucleus sizes, crystallization of liquids or high concentration solutes are preferred so that crystallization peaks are observable in the DSC. The crystallization peak is then associated with the mean droplet size $|\bar{R}|$, since close to this peak most droplets can form stable nuclei and crystallize. To obtain stable polymorphs, solution-crystallization from microemulsions is preferred, as then the substantial supersaturation decrease as the nuclei grow means that large initial supersaturations are required in order to create (near) stable nuclei and these help ensure the nucleation barriers to all polymorphs are surmountable. To ensure crystallization of stable polymorphs from microemulsions, the crystallization should only just be possible, so that the (near) stable nuclei only form in the largest droplets with the highest supersaturations. Hence the initial supersaturations and number of solute molecules in these droplets will be significantly higher than the mean values.

4.4 Crystallization of inorganic systems in microemulsions

Recently we have extended the thermodynamic control of crystallization methodology to inorganic polymorphic systems. There is much literature detailing inorganic nanoparticle formation via the mixed microemulsion approach (Ganguli et al, 2010). However, often the nanoparticles obtained are amorphous and so require high temperature and/or high pressures to introduce crystallinity. For instance, many literature examples concerning the formation of titania nanoparticles produce the crystallinity via subsequent calcining (e.g. Fernández-Garcı et al., 2007; Fresno et al., 2009) or a combined microemulsion-solvothermal process (e.g. Kong et al., 2011). Many methods also involve continual stirring. This is not necessary for microemulsions since they are thermodynamically stable, and stirring may disrupt any potential thermodynamic control if larger transient droplets are formed from multiple colliding droplets. Our microemulsion methodology enables direct crystallization of the nanocrystals of rutile, the stable form of titania, at room temperature provided the solute concentrations are kept sufficiently low and the crystallization is confined (predominantly) to the dispersed phase. To illustrate this, a microemulsion comprising 1.74 g of cyclohexane, 1-hexanol and Triton X-100 in the volume ratio of 7 : 1.2 : 1.8, and 180 µl of 2M HCl as the dispersed phase, was prepared. To this was added a solution of 180 µl of titanium isopropoxide (TIPO) in 1.74 g of the cyclohexane, 1-hexanol and Triton X-100 surfactant solution. The TIPO molecules reacted with the water predominantly at the droplet interface so that the resulting titanium dioxide resided mainly in the droplets. The use of 2M HCl slowed down the production of titanium dioxide, preventing gellation, and allowing the crystallization to proceed under thermodynamic control to give the stable rutile phase. TEM after 12 hours confirmed that the nanoparticles of size ~4 nm were crystalline rutile (see Figure 10a). After 3 days the nanocrystals were ~100 nm (see Figure 10b).

Similar microemulsion compositions with surfactant:aqueous mass ratios of 2.5:1 or less and ≤9% by volume of TIPO, produced rutile nanoparticles of good crystallinity. Indeed calcining these particles at 450 °C for 18 hours led to only a small increase in crystallinity (see Figure 10c). The microemulsions gradually took on a blue tinge over several days due to scattering from the growing rutile particles. The growth of the nanoparticles could be increased by the subsequent addition of water to swell the droplets so that an emulsion formed. Notably, even if the water addition occurred only a few minutes after mixing the TIPO solution with the aqueous HCl microemulsion, good crystallinity rutile particles were still formed. In contrast, when the reaction was carried out in the bulk phase, poor crystallinity/amorphous titania was obtained, demonstrating that the microemulsion stage was crucial for the formation of seed rutile nanocrystals. This general strategy of slowing the reaction rate via limited reactants and/or an appropriate pH range can be used to help introduce, or increase, crystallinity of inorganic nanoparticles obtained from microemulsions.

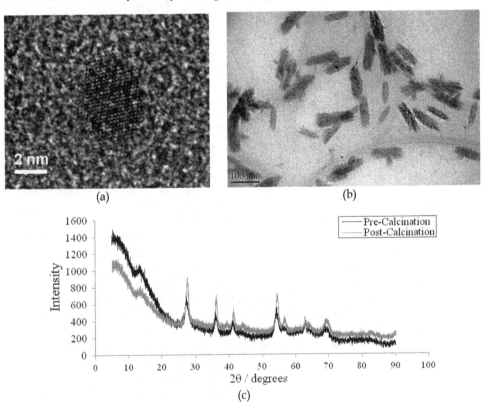

Fig. 10. Crystallization of rutile from microemulsions at room temperature and pressure. (a) High resolution electron microscopy image showing a 4 nm nanocrystal grown after 12 hours. (b) Electron microscopy image of the rutile nanocrystals taken after 3 days. (c) Powder X-ray diffraction trace of the rutile nanocrystals before and after calcination.

4.5 Advantages and drawbacks of crystallization in microemulsions

The use of microemulsions to exert thermodynamic control of crystallization is clearly an advantage whenever stable crystal forms are needed, such as in drug formulations and in obtaining nanocrystals with specific size-dependant properties. However, the much slower growth of crystals in microemulsions, compared to that in bulk solution, may limit industrial applications. A strategy whereby the microemulsion is controllably destabilized once the (near) stable nuclei have formed may circumvent this problem. For instance, the addition of more dispersed phase to swell the droplets into an emulsion may prove advantageous, provided the additional solution results in growth only on the existing (near) stable nuclei and nanocrystals, rather than the nucleation of new crystals, since the latter would be produced under kinetic, rather than thermodynamic, control. An effective approach to ensure this would be to induce the supersaturation of the additional dispersed phase slowly only after the emulsion has formed e.g. by cooling or adding a separate microemulsion containing an antisolvent.

5. Conclusion

Microemulsions present a unique opportunity for both the reliable estimate of critical nucleus sizes and the thermodynamic control of crystallization. The 3D droplet nanoconfinement results in crystallization being limited by the ability to form (near) stable nuclei, rather than critical nuclei, under conditions where crystallization is only just possible. This is a direct consequence of the limited amount of material within a droplet. In solution crystallization there is a substantial supersaturation decrease as a nucleus grows in a nanodroplet. The supersaturation decrease means that very high initial supersaturations are required in a droplet to achieve a (near) stable nucleus, thus enabling nucleation barriers to be readily surmountable. Hence solution crystallization from microemulsions is the only methodology known to-date that can generically crystallize stable polymorphs directly, even when they have insurmountable nucleation barriers in bulk solution. The transient dimer formation in microemulsions provides a mechanism for nuclei growth; we find it is possible to grow crystals ranging from nm to mm in size. Crystallization in microemulsions has already been successfully applied to 'leapfrog' Ostwald's rule of stages and directly crystallize the stable polymorphs of three 'problem' organic systems: glycine, mefenamic acid and ROY. The methodology should be of significant use in the pharmaceutical industry, as it provides the first generic method for finding the most stable polymorph for any given drug, thereby preventing another Ritonavir-type crisis. Microemulsions have also been used to synthesis nanocrystals of rutile without requiring a subsequent calcination step. Other inorganic systems that typically produce amorphous nanoparticulates are also likely to benefit from this approach. Its application to protein crystallization may prove problematic, given the larger size of protein molecules, though droplet clustering to fully encase the protein may occur in these systems alleviating this limitation. Future work will investigate this possibility. A disadvantage of the methodology is that once the (near) stable nuclei are generated, their growth is significantly impeded. Initially this is due to their nanoconfinement, and subsequently, when the nanocrystals grow bigger than these droplets, results from the limited concentration of their molecules in the continuous phase. Controlled microemulsion destabilization strategies, such as adding more of the dispersed phase to form an emulsion, may prove a viable route to circumvent this problem. Finally, given that the ultimate crystal size can vary from nm to mm, depending upon the population of (near) stable nuclei and their subsequent growth rates, there is a significant need for greater understanding of how the growth rates can be tuned. Then the use of microemulsions in crystallization would be truly unrivalled in producing both high crystallinity forms and the desired crystal size.

6. Acknowledgement

We thank EPSRC for funding.

7. References

Allen, K.; Davey, R. J.; Ferrari, E.; Towler, C. & Tiddy, G. (2002). The Crystallization of Glycine Polymorphs from Emulsions, Microemulsions and Lamellar Phases. *Crystal Growth & Design.*, Vol. 2, No. 6 (November, 2002), pp523-527.

Bartell, L. S. (1998). Structure and Transformation: Large Molecular Clusters as Models of Condensed Matter. *Annu. Rev. Phys. Chem.*, Vol. 49, (October 1998), pp43-72.

Becker, R. & Döring, W. (1935). Kinetic Treatment of Germ Formation in Supersaturated Vapour *Ann. Phys.*, Vol. 24, No. 8, (December 1935), pp719-752.

Bowles, R. K.; Reguera D.; Djikaev Y. & Reiss H. (2001). A Theorem for Inhomogeneous Systems: The Generalization of the Nucleation Theorem. *Journal o Chemical Physics*, Vol. 115, (July 2001) pp1853-1866.

Bowles, R. K.; Reguera D.; Djikaev Y. & Reiss H. (2002). Erratum: A Theorem for Inhomogeneous Systems: The Generalization of the Nucleation Theorem. *Journal of Chemical Phyics*, Vol. 116, (February 2002) pp2330.

Chemburkar, S. R.; Bauer, J.; Deming, K.; Spiwek, H.; Patel, K.; Morris, J.; Henry, R.; Spanton, S.; Dziki, W.; Porter, W.; Quick, J.; Bauer, P; Donaubauer, J.; B.; Narayanan, B. A.; Soldani, M.; Riley, D. & McFarland, K. (2000). Dealing with the Impact of Ritonavir Polymorphs on the Late Stages of Bulk Drug Process Development. *Organic Process Research & Development*, Vol. 4, No. 5 (June 2000), pp413-417.

Chen C.; Cook, O.; Nicholson C. E. & Cooper S. J. (2011). Leapfrogging Ostwald's rule of stages: Crystallization of stable γ-glycine directly from microemulsions. *Crystal Growth & Design,* Vol. 11, No. 6 (April 2011), pp2228-2237.

Chew, J. W.; Black, S. N.; Chow, P. S.; Tan, R. B. H. & Carpenter, K. J. (2007). Stable Polymorphs: Difficult to Make and Difficult to Predict. *Crystal Engineering Communications* Vol. 9, No. 2 (January 2007), pp128-130.

Clausse, D.; Babin, L.; Broto, F.; Aguerd, M. & Clausse, M. (1983). Kinetics of Ice Nucleation in Aqueous Emulsions. *Journal of Physical Chemistry* Vol. 87, No. 21 (October 1983), pp4030-4034.

Cooper, S. J.; Nicholson, C. E. & Lui, J. (2008). A simple classical model for predicting onset crystallization temperatures on curved substrates, and its implications for phase transitions in confined volumes. *Journal of Chemical Physics* Vol. 129, No. 12 (September 2008), 124715.

Couchman, P. R. & Jesser, W. A. (1977). Thermodynamic Theory of Size Dependance of Melting Temperature in Metals. *Nature*, Vol. 269, No. 5628, (October 1977), pp481-483.

Denoyel, R. & Pellenq R. J. M. (2002). Simple Phenomenological Models for Phase Transitions in a Confined Geometry. 1: Melting and Solidification in a Cylindrical Pore. *Langmuir*, Vol. 18, No. 7, (April 2002), pp2710-2716.

Erdemir, D.; Lee, A. Y. & Myerson, A. S. (2009). Nucleation of Crystals from Solution: Classical and Two-Step Models. *Accounts of Chemical Research*, Vol. 42, No. 5, (May 2009), pp621-629.

Eriksson, J. C. & Ljunggren, S. (1995). Comments on the Alleged Formation of Bridging Cavities/Bubbles Between Planar Hydrophobic Surfaces. *Langmuir*, Vol. 11, No. 6, (Month 1995), pp1145-1153.

Fletcher N. H. (1958). Size Effect in Heterogeneous Nucleation. *Journal of Chemical Physics*, Vol. 29, No. 3, (May 1958), pp572-576.

Fernández-García, M.; Belver, C.; Hanson, J. C.; Wang, X. & Rodriguez, J. A. (2007). Anatase-TiO2 Nanomaterials: Analysis of Key Parameters Controlling Crystallization. *Journal of the American Chemical Society*, Vol. 129, No. 44, (January 2007), pp13604-13612.

Fresno, F.; Tudela, D.; Coronado, J. M. & Soria J. (2009). Synthesis of Ti1_xSnxO2 nanosized photocatalysts in reverse microemulsions. *Catalysis Today*, Vol. 143, No. 3-4 (May 2009), pp230-236.

Füredi-Milhofer, H.; Garti, N. & Kamyshny, A. (1999). Crystallization from microemulsions - a novel method for the preparation of new crystal forms of aspartame. *Journal of Crystal Growth*, Vol. 198, (March 1999), pp1365-1370.

Ganguli, A. K.; Ganguly A. & Vaidya, S. (2010). Microemulsion-based Synthesis of Nanocrystalline Materials. *Chemical Society Reviews*, Vol. 39, No. 2, (January 2010), pp474-485.

Gibbs, J. W. (1876). On the equilibrium of heterogeneous substances. *Trans. Connect. Acad. Sci.*, Vol. 3, pp108-248.

Gibbs, J. W. (1878). On the equilibrium of heterogeneous substances. *Trans. Connect. Acad. Sci.*, Vol. 16, pp343-524.

Kashchiev, D. (1982). On the Relation Between Nucleation Work, Nucleus Size and Nucleation Rate. *Journal of Chemical Physics*, Vol. 76, No. 10, (May 1982), pp5098-5102.

Jamieson, M. J.; Nicholson, C. E. & Cooper, S. J. (2005). First Study on the Effects of Interfacial Curvature and Additive Interfactial Density on Heterogeneous Nucleation. Ice Crystallization in Oil-in-Water Emulsions and Nanoemulsions with Added 1-heptacosanol. *Crystal Growth & Design*, Vol. 5, No. 2, (March 2005), pp451-459.

Kong, W.; Liu, B.; Ye, B.; Yu, Z.; Wang, H.; Qian, G. & Wang, Z. (2011) An Experimental Study on the Shape Changes of TiO2 Nanocrystals Synthesized by Microemulsion-Solvothermal Method, *Journal of Nanomaterials* Volume 2011, Article ID 467083, 6 pages.

Liu, J.; Nicholson, C. E. & Cooper, S. J. (2007). Direct measurement of critical nucleus size in confined volumes. *Langmuir*, Vol. 23, No. 13, (June 2007), pp7286-7292.

Nicholson, C. E.; Chen, C.; Mendis, B. & Cooper, S. J. (2011) 'Stable Polymorphs Crystallized Directly under Thermodynamic Control in Three-Dimensional Nanoconfinement: A Generic Methodology' *Crystal Growth & Design*, Vol 11, No. 2 (February 2011), pp. 363-366.

Nicholson, C. E. & Cooper, S. J. (2011). Crystallization of Mefenamic Acid from DMF Microemulsions: Obtaining Thermodynamic Control through 3D Nanoconfinement. *Crystals* Vol. 1, No. 3, (September 2011), pp. 195-205.

Nicholson, C. E.; Cooper, S. J.; Marcellin, C. & Jamieson, M. J. (2005). The Use of Phase-Inverting Emulsions to Show the Phenomenon of Interfacial Crystallization on both Heating and Cooling. *Journal of the American Chemical Society*, Vol. 127, No. 34, (August 2005), pp11894-11895.

Ostwald, W. Z. (1897). Studies of the Formation and Transformation of solid Substances. *Zeitschrift für Physikalische Chemie*, Vol. 22, (1897), pp289-330.

Oxtoby, D. W. & Kashchiev, D. (1994). A General Relation Between the Nucleation Work and the Size of the Nucleus in Multicomponent Nucleation. *Journal of Chemical Physics*, Vol. 100, No. 10, (May 1994), pp7665-7671.

Petit, C.; Lixon, P. & Pileni, M. P. (1990). Synthesis of Cadmium Sulfide in situ in Reverse Micelles. 2: Influence of the Interface on the Growth of the Particles. *Journal of Physical Chemisty*, Vol. 94, No. 4, (February 1990), pp1598–1603.

Petrov, O. & Furó, I. (2006). Curvature-dependant Metastability of the Solid Phase and the Freezing-Melting Hysteresis in Pores. *Phyical Reiew E*, Vol. 73, No. 1, (January 2006), 001608-1 – 001606-7.

Popovitz-Biro, R.; Wang, J. L.; Majewski, J.; Shavit, E.; Leiserowitz, L.; Lahav, M. (1994). Induced Freezing of Supercooled Water into Ice by Self-Assembled Crystalline Monolayers of Amphiphilic Alcohols at the Air-Water Interface. *Journal of the American Chemical Society*, Vol. 116, No. 4, (February 1994), pp1179-1191.

Pruppacher, H. R. (1995). A New Look at Homogeneous Ice Nucleation in Supercooled Water Drops. *Journal of Atmospheric Science*, Vol. 52, No. 11, (June 1995), pp1924-1933.

Reguera, D.; Bowles, R. K.; Djikaev Y. & Reiss, H. (2003). Phase Transitions In Systems Small Enough to be Clusters. *Journal of Chemical Physics*, Vol. 118, No. 1, (January 2003), pp340-353.

Speedy, R. J. (1987). Thermodynamic Properties of Supercooled Water at 1 atm. *Journal of Phyical Chemistry*, Vol. 91, No. 12, (June 1987), pp3354-3358.

Vanfleet, R. R. & Mochel, J. M. (1995). Thermodynamics of Melting and Freezing in Small Particles. *Surf. Sci.*, Vol. 341, No. 1-2, (November 1995), pp40-50.

Vekilov, P. G. (2010). Nucleation. *Crystal Growth & Design*, Vol. 10, No. 12, (December 2010), pp5007-5019.

Volmer, M, & Weber, A. (1926). Germ Formation in Oversaturated Figures. *Zeitschrift fürPhysikalische Chemie*, Vol. 119, No. 3-4, (March 1926), pp277-301.

Wood, G. R. & Walton, A. G. (1970). Homogeneous Nucleation Kinetics of Ice from Water. *Journal of Applied Physics*, Vol. 41, No. 7 (June 1970), pp3027-3036.

Xu, D. & Johnson, W. L. (2005). Geometric Model for the Critical-Value Problem of Nucleation Phenomena Containing the Size Effect of Nucleating Agent. *Physical Review B*, Vol. 72, No. 5, (August 2005), pp052101.

Yano, J.; Füredi-Milhofer, H.; Wachtel, E. & Garti, N. (2000). Crystallization of Organic Compounds in Reversed Micelles. 1: Solubilization of Amino Acids in Water-

Isooctane-AOT Microemulsions. *Langmuir*, Vol. 16, No. 26, (December 2000), pp10005-10014.

Zarur, A. J. & Ying, J.Y. (2000). Reverse Microemulsion Synthesis of Nanostructured Complex Oxides for Catalytic Combustion. *Nature*, Vol. 403, No. 6765, (January 2000), pp65–67.

7

Chemical, Physicochemical and Crystal – Chemical Aspects of Crystallization from Aqueous Solutions as a Method of Purification

Marek Smolik

Faculty of Chemistry, Silesian University of Technology, Gliwice
Poland

1. Introduction

This chapter is intended to discuss the effect of chemism of crystallizing and co-crystallizing substances (i.e., their chemical, physicochemical and crystal–chemical properties), as well as some other factors on efficiency of their separation and purification during crystallization from aqueous solutions.

There are three main aims of crystallization (Rojkowski & Synowiec, 1991): creation of the solid phase, forming crystals, purification of substances.

While using crystallization for purification and separation of various substances, as well as for enrichment of trace amounts of new–found radioactive elements, it was established that (in addition to many others) chemical factors strongly affected the mentioned operations. Mechanisms of trace radioactive elements' co-crystallization and the significance of these factors on their enrichment efficiency were reviewed in some works (Przytycka, 1968; Niesmeanov, 1975).

The influence of factors determining the structures of salts of crystal ionic lattices, (salts - considered as ionic coordination compounds) and their ability to isomorphous and isodimorphous mixing on their possibility to crystallization separation was thoroughly discussed by Balarew (1987). Whereas developments concerning inclusions of isomorphous impurities during crystallization from solutions were reviewed by Kirkova et al. (1998).

The discovered settlements were useful in the preliminary assessment of the effectivity of the crystallization method for purification of substances (Kirkova, 1994), for concentration microimpurities (Zolotov & Kuzmin, 1982), for growing of single crystals of specific properties (Byrappa, et al., 1986; Demirskaya, et al. 1989), as well as in explanation of the genesis of some minerals (Borneman-Starinkevich, 1975).

In spite of development of solvent extraction and ionic exchange methods, crystallization is still a very attractive method of purification, particularly in the preparation of numerous high–purity inorganic substances (HPIS). There are two main reasons for that:

- for crystallization purification of a substance only the simplest reagents are necessary (like water or other solvents, sometimes salting out or complexing agends), which can be easily purified to the level suitable for HPIS and readily removed after crystallization;

- in the case of many HPIS (especially crystalline preparations) crystallization is often the final stage of their preparation, which can be simply carried out without incidental contamination.

2. Crystallization as a method of purification

The crystallization should, in principle, yield very significant purification of a substance, but for the phenomenon of transport of accompanying impurities into the crystal, which may happen in the following ways presented in the simplified scheme below.

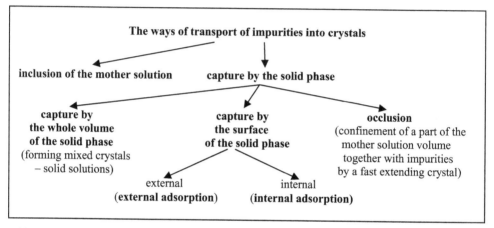

Although a suitable choice of crystallization conditions (supersaturation, rate of crystallization), as well as the ways of separation of crystals from mother solutions (filtration, washing) permits minimizing the capture of impurities derived from the inclusion of the mother solution, occlusion or external adsorption, it is impossible to restrict impurities originated from the capture by the whole volume of the solid phase or internal adsorption.[1]

The highest effect of purification may be expected when impurities are not captured by the solid phase of crystals but get into crystals as a result of the mother solution's residue, which cannot be removed by filtration[2]. In this boundary case the efficiency of crystal purification after its separation from the mother solution (without washing) is defined by the equation (Gorshtein, 1969):

$$\frac{1}{K_k} = \frac{e'_k}{e'_o} = \frac{100 - C_k}{C_k} \cdot \frac{C_r}{100 - C_r} \cdot \frac{1}{1 - \alpha} \tag{1}$$

---

[1] Internal adsorption takes place when microcomponents cannot form solid solutions with macrocomponent (Niesmieanov,1975)], it is a rather sparsely occurring phenomenon (Przytycka,1968).
[2] During the crystallization without stirring or with not vigorous stirring, big, aggregated crystals (twins, intergrowth) are obtained. The presence of cavities on their surface obstructs the separation of mother solution from these crystals, which results in lowering their purity. Stirring during the crystallization at a considerable concentration of crystals causes rounding of crystals because of abrasion. Then large crystals adopt the form of spheres or ellipsoids, whose separation from mother solution by means of filtration is easier, which leads to higher purity of final product (Matusevich, 1961; Bamforth, 1965).

Chemical, Physicochemical and Crystal –
Chemical Aspects of Crystallization from Aqueous Solutions as a Method of Purification

173

where:

K_k– crystal purification coefficient (multiplicity of lowering initial microcomponent contents in crystal)

e'_k– relative contents of a microcomponent in the crystal [ppm],

e'_o– initial relative contents of a microcomponent (before crystallization) [ppm],

C_k– contents of the macrocomponent in the crystal [%], (100–C_k) – crystal humidity [%],

C_r– contents of the macrocomponent in the mother solution, (its solubility) [%],

α– degree of crystallization of the macrocomponent.

However, in reality, in numerous cases a microcomponent is captured by the solid phase, mainly by forming mixed crystals. Micro and macrocomponents form real mixed crystals (solid solutions) if they are isomorphous or isodimorphous.

2.1 Co–crystallization coefficients

2.1.1 Homogeneous distribution coefficient $D_{2/1}$ (Henderson– Kraček, Chlopin)

Homogeneous partition takes place in equilibrium conditions between the whole mass of a crystal and the mother solution, and is described by the Chlopin equation:

$$\frac{n\rho_s}{m_s} = K_X \frac{(n_o - n)\rho_r}{m_r} \tag{2}$$

where : n – number of moles of the microcomponent in a crystal, n_o – the whole number of moles of the microcomponent in the system, m_s – the mass of a crystal, m_r – the mass of solution, ρ_s – density of crystal, ρ_r – density of solution, K_X – Chlopin constant.

Taking into account that $\dfrac{n\rho_s}{m_s} = \dfrac{n}{\dfrac{m_s}{\rho_s}} = \dfrac{n}{V_s} = C_s$ as well as $\dfrac{(n_o - n)\rho_r}{m_r} = \dfrac{(n_o - n)}{m_r} = \dfrac{(n_o - n)}{V_r} = C_r$

where: V_s and V_r – volumes of the solid phase and the solution, C_s and C_r – concentration of a microcomponent in the solid phase and in the solution it is possible to obtain an equation, identical to the well-known Berthelot–Nernst equation describing the partition of a substance between two immiscible solvents $K_x = C_s/C_r$.

During the crystallization from the solution containing two components: macrocomponent (1) and microcomponent (2) the ratio of their partition coefficients defines the equilibrium co-crystallization coefficient:

$$D = D_{2/1} = \frac{K_2}{K_1} = \frac{\left(\dfrac{C_s}{C_r}\right)_2}{\left(\dfrac{C_s}{C_r}\right)_1} = \frac{\left(\dfrac{C_2}{C_1}\right)_s}{\left(\dfrac{C_2}{C_1}\right)_r} = \frac{C_{2s} \cdot C_{1r}}{C_{1s} \cdot C_{2r}}. \tag{3}$$

Substituting: n_o and n – number of moles of microcomponent in the whole system and in the solid phase, and z_o and z – number of moles of macrocomponent in the whole system and in the solid phase into equation (3) and suitable rearranging, it is possible to obtain a more convenient Henderson & Kraček equation (Niesmieanov, 1975):

$$D_{2/1} = \frac{n(z_o - z)}{z(n_o - n)} \tag{4}$$

Further transformation of this equation gives other, often used practical formulae (Smolik, 2004):

$$D_{2/1} = \frac{\dfrac{n}{z}}{\dfrac{(n_o - n)}{(z_o - z)}} = \frac{e'_s}{e'_r} \tag{5}$$

and

$$D_{2/1} = \frac{\dfrac{n}{n_o}\left(\dfrac{z_o}{z_o} - \dfrac{z}{z_o}\right)}{\dfrac{z}{z_o}\left(\dfrac{n_o}{n_o} - \dfrac{n}{n_o}\right)} = \frac{\beta(1-\alpha)}{\alpha(1-\beta)}, \tag{6}$$

where $\alpha = z/z_o$ is the degree of crystallization of macrocomponent, $\beta = n/n_o$ is the degree of co-crystallization of microcomponent, e'_s and e'_r are relative concentrations of microcomponent in the solid phase and in the mother solution, respectively ([ppm] in relation to macrocomponent), $D_{2/1}$ – homogeneous partition coefficient (co-crystallization coefficient).

2.1.2 Heterogeneous (logarithmic) distribution coefficients λ (Doerner–Hoskins)

Logarithmic partition can take place if the equilibrium between the whole mass of crystal does not exist, but only between the surface layer of a crystal and solution. If $D_{2/1} \neq 1$, the concentration of microcomponent in the solution during the crystallization will be changing continuously. So the microcomponent will distribute in the crystal in a stratified manner ("onion" structure). This process for the elementary layer of the crystal may be described by the equation parallel to that of Henderson–Kraček: $\dfrac{dn}{dz} = \lambda \dfrac{(n_o - n)}{(z_o - z)}$,

where the meaning of n_o, n, z_o, z is the same as previously described and λ is the heterogeneous (logarithmic) partition coefficient. The integration of this expression yields the known equations (Doerner & Hoskins, 1925):

$$\ln\frac{n_o}{n_o - n} = \lambda \ln\frac{z_o}{z_o - z}, \tag{7}$$

or

$$\lambda = \frac{\log(1-\beta)}{\log(1-\alpha)} \tag{8}$$

Chemical, Physicochemical and Crystal –
Chemical Aspects of Crystallization from Aqueous Solutions as a Method of Purification

175

Both homogeneous and heterogeneous partitions are boundary cases of distribution of the microcomponent between the solid phase and the mother solution. Experimental study involving which of both coefficients retains constant value with the increase of the degree of crystallization gives information on what partition is actually taking place.

2.2 Homogeneous distribution coefficients $D_{2/1}$ as indicators of crystallization efficiency

Homogeneous partition coefficients $D_{2/1}$ are a convenient measure of crystallization efficiency as a method of purification. In the case of homogeneous partition of microcomponent in the solid phase, the final result of purification (without washing) may be expressed by the formula (Gorshtein, 1969), derived on the basis of the balance of amounts of the microcomponent during the crystallization:

$$\frac{1}{K_k} = \frac{e_k'}{e_o'} = \frac{D_{2/1}'}{\alpha\, D_{2/1}'+1-\alpha} + \frac{100-C_k}{C_k} \cdot \frac{C_r}{100-C_r} \cdot \frac{1-D_{2/1}'}{\alpha\, D_{2/1}'+1-\alpha} \tag{9}$$

After careful washing of the crystals by pure, saturated solution of the macrocomponent, inclusions of the mother solution, as well as microcomponents adsorbed on the surface of the crystal, will be removed. The result of the purification in this case will be:

$$\frac{1}{K_k} = \frac{e_{k(p)}'}{e_o'} = \frac{D_{2/1}}{\alpha D_{2/1}+1-\alpha} \tag{10}$$

where: K_K – multiplicity of lowering initial microcomponent contents (e_0', $e_{k(p)}'$ - initial contents of microcomponent –[ppm] in crystal and in washed crystal after crystallization); $D_{2/1}(D_{2/1}') = e_s(e_{k(p)}')/e_r'$ (e_s' and e_r' – contents of micomponent - [ppm] in the solid phase and the mother solution); $D_{2/1}$ – isomorphous co-crystallization coefficient of microcomponent, $D_{2/1}'$ – adsorption–isomorphous co-crystallization coefficient of microcomponent; expression ($D_{2/1}'$ – $D_{2/1})/D_{2/1}'$ qualifies a relative importance of adsorption in the capture of microcomponent by the solid phase (Gorshtein, 1969).

Knowing $D_{2/1}(D_{2/1}')$ and using the equations (9) and (10) it is possible to evaluate a number of crystallizations in the conditions of homogeneous partition of the microcomponent (at $D_{2/1}(D_{2/1}')$ =const.) necessary to achieve a desirable degree of purification of crystals. An example of such evaluation for $NiSO_4 \cdot 7H_2O$ is presented in Table 1.

$D_{2/1}'$	$D_{2/1}$	Number of crystallizations (k)	The whole yield of purification $(m_f/m_0) \cdot 100\% = (\alpha^k) \cdot 100\%$
0.75	0.75	15	0.003
0.50	0.50	6	1.56
0.25	0.25	3	12.5
0.10	0.10	2	25.0
0.05	0.05	1	50.0

Table 1. The number of $NiSO_4 \cdot 7H_2O$ crystallizations necessary to achieve 10–fold lowering of its initial contents of microcomponent for various levels of coefficient $D_{2/1}$ ($D_{2/1}'$) (Smolik, 2004), $C_k=98\%$, $C_r=50\%$, $\alpha=0,50$ (50%), $m_0(m_f)$ – initial (after k crystallizations) mass of crystals

The data presented in Table 1 show that the level of $D_{2/1}$ ($D_{2/1}'$) is a very important parameter for the evaluation of crystallization efficiency as a method of purification, and therefore, its knowledge is significant in planning the utilization of crystallization in different stages of preparation of high purity substances.

2.3 Practical and equilibrium $D_{2/1}$ coefficients

However, crystallization processes are usually realized at non–equilibrium conditions and the obtained, in this case practical (effective), partition coefficients ($D_{P2/1}$) depend on the ways in which the crystallization is carried out. This dependence may be presented by the following expression (Kirkova et al., 1996):

$$D_{P2/1} = D_{\circ 2/1}\, \Theta(T, \varsigma, \omega, m_2, m_j\, \kappa) \tag{11}$$

where $D_{\circ 2/1}$ – equilibrium co-crystallization coefficient, Θ - imbalance factor, which is a function of temperature(T), supersaturation of the solution (ς), rate of stirring (ω), concentration of microcomponent 2 (m_2), concentration of other microcomponents (m_j) and other factors (κ).

Since the equilibrium co-crystallization coefficient does not depend on crystallization conditions, it may be compared with various crystallization systems. Hence, it is important to trace the determination methods of such coefficients.

2.4 Methods of determination of equilibrium distribution coefficients

The possibility of achieving equilibrium homogeneous partition of microcomponents between solution and crystal solid phase was proved by Chlopin. Subsequent investigations in this area (Gorshtein, 1969; Chlopin, 1957; Zhelnin & Gorshtein, 1971) elaboration upon several methods to accomplish equilibrium partition of microcomponents in the crystal, among which the method of isothermal decreasing of supersaturation and the method of long–time stirring of crushed crystals in their saturated solution are most often used.

2.4.1 The method of isothermal decreasing of supersaturation

It relies on cooling a saturated solution without stirring to the end temperature of crystallization (so that no crystal would appear) and after that on vigorous stirring at the constant end temperature until a complete removal of supersaturation takes place (usually for 3 - 360 h) (Zhelnin & Gorshtein, 1971; Chlopin, 1957). An example of such determination of $D_{2/1}{}^{eq}$ is presented in Fig. 1

2.4.2 The method of long–time stirring of crushed crystals in their saturated solution

The equilibrium is reached starting either from the initial concentration ratio of a microcomponent in crystal and in solution exceeding the expected value of its equilibrium coefficient ($D_{\circ MAX}=e'_{ko}("contaminated"\ crystal)/e'_{ro}("purified"\ solution)$) – achieving equilibrium "from above" or from this ratio lower than the expected value mentioned above ($D_{\circ min}=e''_{ko}("purified\ crystal")/e''_{ro}("contaminated"\ solution)$) – achieving equilibrium "from below" (Fig. 2).

Chemical, Physicochemical and Crystal –
Chemical Aspects of Crystallization from Aqueous Solutions as a Method of Purification

177

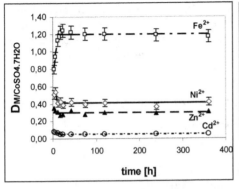 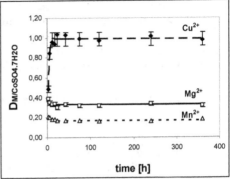

Fig. 1. Changes of co-crystallization coefficients, $D_{2/1}$ of M^{2+} ions as the effect of isothermal levelling of supersaturation during the crystallization of $CoSO_4 \cdot 7H_2O$ at 20 ºC (Smolik, 2003)

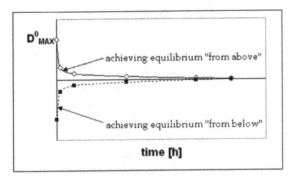

Fig. 2. The principle of the long–time stirring method for the determination of the equilibrium coefficients $D_{2/1}$ (Zhelnin & Gorshtein, 1971; Chlopin, 1957)

When selecting values D^o_{max} and D^o_{min} the highest and the lowest values of $D_{2/1}$ obtained during crystallization by the first method are usually taken into consideration. The experiments are carried out in the following way:

Micro-compo-nent M^{2+}	Initial $D_{2/1}$		Average $D_{2/1}$ after long time stirring		Average equilibrium $D_{2/1}$
			for D^o_{min}	for D^o_{MAX}	
	D^o_{min}	D^o_{MAX}	$\bar{D} \pm t_\alpha \dfrac{s}{\sqrt{n}}$	$\bar{D} \pm t_\alpha \dfrac{s}{\sqrt{n}}$	$\bar{D} \pm t_\alpha \dfrac{s}{\sqrt{n}}$
Ni^{2+}	1.22	1.86	1.52 ± 0.06	1.62 ± 0.09	1.57 ± 0.06
Cu^{2+}	0.08	0.22	0.14 ± 0.02	0.15 ± 0.01	0.14 ± 0.01
Co^{2+}	0.77	1.89	1.18 ± 0.09	1.20 ± 0.05	1.19 ± 0.04
Fe^{2+}	0.42	1.44	0.72 ± 0.05	0.79 ± 0.06	0.76 ± 0.04
Mg^{2+}	0.16	3.47	1.33 ± 0.12	1.40 ± 0.09	1.36 ± 0.07
Mn^{2+}	0.13	0.28	0.17 ± 0.02	0.20 ± 0.02	0.19 ± 0.02

Table 2. Determination of equilibrium $D_{2/1}$ coefficients of M^{2+} ions during the crystallization of $ZnSO_4 \cdot 7H_2O$ at 23ºC (Smolik, 2000a)

Crushed "contaminated" crystals (crushed to pass a 0.1mm sieve – ϕ<0.1mm) are introduced into several beakers together with their saturated "purified" solution. Crushed "purified" crystals (crushed to pass a 0.1mm sieve – ϕ<0.1mm) and their "contaminated" saturated solution are introduced to some other beakers. Contents of the beakers are stirred for ~360 h with a magnetic stirrer at constant temperature (**Table 2**).

3. Thermodynamic approach to the calculation of equilibrium $D_{2/1}$ coefficients

The possibility of achieving a thermodynamic equilibrium during crystallization from solutions, as well as melts, as proved by Chlopin (1957), permits introducing a thermodynamic partition coefficient. Substituting concentrations of microcomponent (2) and macrocomponent (1) in equation (3) with their activities (a_{1S}, a_{2s}, a_{1r}, a_{2r}) it is possible to obtain an expression for thermodynamic co-crystallization coefficient $D^o_{2/1}$. (Kirkova et al., 1996; Ratner, 1933).

$$D^o_{2/1} = \frac{a_{2s} \cdot a_{1r}}{a_{2r} \cdot a_{1s}} = \exp\left(-\frac{\Delta\mu^o_{2/1}}{RT}\right) \tag{12}$$

$\Delta\mu^o_{2/1} = \Delta\mu^o_2 - \Delta\mu^o_1$, where $\Delta\mu^o_2$ and $\Delta\mu^o_1$ are the changes of standard molar chemical potential of components (2) and (1) respectively during the transition from the liquid phase (r) into the solid phase (s). Two cases should be distinguished here: 1) substance (2) is not isomorphous with substance (1), i.e., it crystallizes in different crystal systems (different space groups); 2) substance (2) is isomorphous with substance (1).

In the first case the formation of mixed crystals by substance (2) with substance (1) may indicate the existence (besides the basic form [II] of microcomponent [2]) of a polymorphous form (I), metastable in suitable conditions, which is isomorphous with the crystal of the host: macrocomponent – substance (1). The transition of the substance (2) of the structure (II) into its metastable form of the structure (I) is connected with the increase of chemical potential: $\Delta\mu^o_{II\to I} = \mu^{o,I}_{2s} - \mu^{o,II}_{2s}$. Then $\Delta\mu^o_2 = \mu^{o,I}_{2s} - \mu^o_{2r} = \mu^{o,I}_{2s} - \mu^{o,II}_{2s} + \mu^{o,II}_{2s} - \mu^o_{2r} = \Delta\mu^o_{II\to I} + \mu^{o,II}_{2s} - \mu^o_{2r}$, where $\Delta\mu^o_{II\to I} = \mu^{o,I}_{2s} - \mu^{o,II}_{2s}$ – free partial molar enthalpy of phase transition II→I of crystals of microcomponent (2) of structure (II) into the structure (I) proper to that of macrocomponent(1). Therefore (Smolik, 2004):

$$D^o_{2,II/1,I} = \frac{a_{1r}a_{2s}}{a_{1s}a_{2r}} = \frac{a_{1r}a^I_{2s}}{a_{1s}a_{2r}} = \exp\left(-\frac{\Delta\mu^o_2 - \Delta\mu^o_1}{RT}\right) = \exp\left(-\frac{\Delta\mu^o_{II\to I} + \mu^{o,II}_{2s} - \mu^o_{2r} + \mu^o_{1r} - \mu^o_{1s}}{RT}\right) =$$
$$= \exp\left(\frac{\mu^o_{1s} - \mu^o_{1r}}{RT}\right) \cdot \exp\left(\frac{\mu^o_{2r} - \mu^{o,II}_{2s}}{RT}\right) \cdot \exp\left(-\frac{\Delta\mu^o_{II\to I}}{RT}\right) = \frac{a^o_{1r}}{a^o_{2r}} \cdot \exp\left(-\frac{\Delta\mu^o_{II\to I}}{RT}\right) \tag{13}$$

If two double salts: $B_bE_eL_l$ (1) i $B_bE_eL_l$ (2) capable of forming solid solutions by the exchange of B ions into B' ones (it is possible to exchange E ions into E' or L into L') dissociate into ions in aqueous solution according to the reaction:

$$B_bE_eL_l \leftrightarrows bB^{(z1)+} + eE^{(z2)+} + lL^{(z3)-}$$

and

$$B'_bE_eL_l \leftrightarrows bB'^{(z1)+} + eE^{(z2)+} + lL^{(z3)-}$$

(obviously: $bz_1 + ez_2 - lz_3 = 0$) the following general formula may be derived for thermodynamic co-crystallization coefficient $D^o_{2/1}$ (Balarew, 1987; Smolik & Kowalik, 2010):

$$D_{2/1} = \left(\frac{x_2 \cdot m_1}{x_1 \cdot m_2} \right) = \left(\frac{m_{01} \cdot \gamma_{m01}}{m_{02} \cdot \gamma_{m02}} \right)^{\frac{v}{b}} \cdot \left(\frac{\gamma_{m2}}{\gamma_{m1}} \right)^{\frac{v}{b}} \cdot \frac{f_1}{f_2} \cdot \exp\left(-\frac{\Delta\mu^o_{II \to I}}{bRT} \right) \tag{14}$$

where: $m_{01}(m_{02})$, $\gamma_{m01}(\gamma_{m02})$ – molal solubility ([mol/kg]) of the salt $B_bE_eL_l(B'_bE_eL_l)$) and mean molal activity coefficient of the salt $B_bE_eL_l(B'_bE_eL_l)$ in its binary saturated solution; $m_1(m_2)$, $\gamma_{m1}(\gamma_{m2})$ – molality and mean molal activity coefficient of the salt $B_bE_eL_l(B'_bE_eL_l)$ in the ternary solution being in equilibrium with $B_b(B'_b)E_eL_l$ solid solution; $x_1(x_2)$ – mole fraction of $B(B')$ ion and $f_1(f_2)$ – activity coefficient of ion $B(B')$ in this solid solution; $\Delta\mu^o_{II \to I}$ – the partial molar Gibbs free energy of the phase transition of the salt $B'_bE_eL_l$ from its structure (II) into the structure (I) of the salt $B_bE_eL_l$, $v = b + e + l$, R – gas constant, T – temperature [K]

In the other case involving isomorphous substances (1) and (2) $\Delta\mu^o_{I \to I} = 0$. Hence:

$$D_{2/1} = \left(\frac{m_{01} \cdot \gamma_{m01}}{m_{02} \cdot \gamma_{m02}} \right)^{\frac{v}{b}} \cdot \left(\frac{\gamma_{m2}}{\gamma_{m1}} \right)^{\frac{v}{b}} \cdot \frac{f_1}{f_2} \tag{15}$$

Equations (14) and (15) should, in principle, permit calculating exactly the equilibrium partition coefficient $D_{2/1}$, if molal solubilities and all activity coefficients (in the aqueous and the solid phases), as well as the partial molar Gibbs free energy of the phase transition, were known. However, these data (except for molal solubilities) are rarely available. In contrast to mean molal activity coefficients in binary saturated solutions ($\gamma_{m01}, \gamma_{m02}$), as well as those in the ternary solution being in equilibrium with $B_b(B'_b)E_eL_l$ solid solution (γ_{m1}, γ_{m2}) which are sometimes directly accessible or calculable by means of Pitzer equations, activity coefficients in the solid solution (f_1, f_2), as well as the partial molar free energy of the phase transition, are generally unknown (except for very rare individual cases of crystallization systems: macrocomponent(1) – microcomponent(2)).

The attempts to estimate $D_{2/1}$ coefficients by means of simplified equations (taking into account only activity coefficients in the liquid phase) are connected with huge errors, which proves that they result from the lack of the activity coefficients in the solid solution (f_1, f_2) as well as the partial molar Gibbs free energy of the phase transition $\Delta\mu^o_{II \to I}$ (Smolik, 2004).

$$\frac{\Delta[\%]}{100} = \frac{|D_{exp.} - D_{cal.}|}{D_{exp.}} = \left| 1 - \frac{D_{cal.}}{D_{exp.}} \right| = \left| 1 - \frac{f_2}{f_1} \cdot \exp\left(\frac{\Delta\mu^0_{II \to I}}{bRT} \right) \right| \tag{16}$$

In the case where coefficients $D_{2/1}$ are independent of mixed crystal composition, the ratio f_2/f_1 remains constant (Balarew, 1987). Assuming the regular solution approximation this ratio may be expressed by the following equation:

$$\frac{f_1}{f_2} = \exp\left(\frac{\overline{\Delta H_1} - \overline{\Delta H_2}}{RT} \right) \tag{17}$$

where $\Delta \overline{H}_1 - \Delta \overline{H}_2$ is the difference in the partial molar enthalpies of mixing.

According to Balarew (1987) this is the result of the difference in coordination environment around the two substituting ions, affected by ionic size differences ($\Delta r/r$), metal – ligand bond energy differences ($\Delta \varepsilon$) with respect to the enthalpy of mixing (Urusov, 1977), as well as the difference in the energy determined by the crystal field (in the case non Jahn–Teller ions):

$$\Delta \overline{H}_1 - \Delta \overline{H}_2 = w_1 \cdot f\left(\frac{\Delta r}{r}\right) + w_2 \cdot \varphi(\Delta \varepsilon) + w_3 \cdot \psi(\Delta s) + ... \tag{18}$$

Hence:

$$D_{2/1} = \left(\frac{c_{01} \cdot \gamma_{c01}}{c_{02} \cdot \gamma_{c02}}\right)^{\frac{v}{b}} \cdot \left(\frac{\gamma_{c2}}{\gamma_{c1}}\right)^{\frac{v}{b}} \cdot \exp\left(-\frac{\Delta \mu_{II \to I}}{RT}\right) \cdot \exp \frac{w_1 \cdot f\left(\frac{\Delta r}{r}\right) + w_2 \cdot \varphi(\Delta \varepsilon) + w_3 \cdot \psi(\Delta s) + ...}{bRT} \tag{19}$$

where: f, φ, ψ, ... – functions sought for, w_1, w_2, w_3, ... – estimated coefficients.

To derive an equation for estimating $D_{2/1}$ by finding the functions (f, φ, ψ, ...) and coefficients (w_1, w_2, w_3, ...), it is necessary to check how $D_{2/1}$ coefficients depend on various factors.

4. The dependence of co–crystallization coefficients, $D_{2/1}$ on chemical, physicochemical and crystal–chemical properties of co–crystallizing salts and ions

Equilibrium co-crystallization coefficients are determined in the conditions ensuring that they do not depend on hydrodynamic and kinetic conditions of crystallization. However, they are affected by several factors both "external" (in relation to the co-crystallizing substances) and "internal" (resulting from chemical, physicochemical and crystal–chemical properties of the co-crystallizing substances).

"External" factors have chemical characteristics (kind and composition of the solvent – the liquid phase, the presence of ions or other foreign substances, the presence of complexing agents, acidity (pH) of solution, from which crystallization takes place) or non–chemical ones (e.g., temperature). "Internal" factors are presented in Table **3**.

Chemical, physicochemical and crystal–chemical properties of	
Co-crystallizing salts	Co-crystallizing ions
Solubility in water (m_0)	Charge of cation
Crystal system (CS)	Geometrical factor (ionic radius) (r)
Number of molecules of crystallization water (n)	Character of chemical bond (electronegativity) (ε)
Reciprocal solubility in the solid phase ($C^s{}_{MAX}$)	Electronic configuration
The volume of one formal molecule of salt (η^3)	Crystal field stabilization energy (CFSE)
	Cation hardness (h)

Table 3. Chemical, physicochemical and crystal–chemical properties of co-crystallizing salts and co-crystallizing ions

To analyse the influence of the above mentioned factors on co-crystallization coefficients $D_{2/1}$, it is convenient to use correlation coefficients (ρ_{xy}) in the case of the properties that can be formulated quantitatively. For other properties (qualitative), mean $D_{2/1}$ values for salts revealing and not revealing may be compared. On the other hand, co-crystallization coefficients may be considered as a measure of mutual solubility of co-crystallizing salts in the solid phase. The famous Latin rule: "Similia similibus solvuntur" (similar substances will dissolve similar substances) may be useful in the prediction of this solubility and the evaluation of $D_{2/1}$ level.

4.1 Chemical, physicochemical and crystal–chemical properties of co-crystallizing salts

4.1.1 Solubility in water (m_0)

This is the most important factor affecting $D_{2/1}$coefficients. For the co-crystallization of isomorphous salts $(\Delta\mu^0_{II \to I} = 0)$ forming ideal solid and liquid solutions $((\gamma_{m01}/\gamma_{m02})^{(v/b)} \cdot (\gamma_{m2}/\gamma_{m1})^{(v/b)} (f_1/f_2)=1))$ they are expressed by $D_{2/1}=(m_{01}/m_{02})^{v/b}$.

However, this equation is proved true only for non-numerous salts fulfilling the additivity rule (Balarew, 1987). This simplified equation is the basis of the Ruff rule (Ruff et al., 1928):

If $m_{01} > m_{02} \to (m_{01}/m_{02})^{v/b} = D_{2/1} > 1$. (During crystallization of two components, the less soluble one grows rich in crystal).

As it can be seen in Table **4**, despite its simplicity and obviousness, this qualitative rule is not always fulfilled.

Kind of co-crystallizing salts	Number of considered crystallization systems	Crystallization systems fulfilling the Ruff rule [%]
$MSO_4 \cdot nH_2O$	100	62
$MCl_2 \cdot nH_2O$	23	74
$M(HCOO)_2 \cdot 2H_2O$	37	95
Alums, M(III)	9	100
$M^I_2M^{II}(SO_4)_2 \cdot 6H_2O$ M(I), M(II)	59	96
$M^I_2SO_4$	24	54
MX	16	69
$MClO_3, MClO_4, MNO_3, M_2CrO_4$	17	59
mean		72

Table 4. The degree of fulfilling the Ruff rule in some crystallization systems: macrocomponent – microcomponent (Smolik, 2004)

Mean molal activity coefficients of some isomorphous double salts, forming ideal solid solutions ($M^I_2M^{II}(SO_4)_2 6H_2O$ or $M^IM^{III}(SO_4)_2 12H_2O$) in their binary saturated solutions are inversely proportional to the square root of their molal solubility (Hill et al., 1940). Therefore:

$$D_{2/1}=\left(\frac{m_{01} \cdot \gamma_{m01}}{m_{02} \cdot \gamma_{m02}}\right)^{\frac{v}{b}} = \left(\frac{m_{01} \cdot \sqrt{m_{02}}}{m_{02} \cdot \sqrt{m_{01}}}\right)^{\frac{v}{b}} = \left(\frac{m_{o1}}{m_{o2}}\right)^{\frac{v}{2b}} \qquad (20)$$

For double salt dissociating: $NiSO_4 \cdot M^I_2SO_4 \cdot 6H_2O \leftrightarrows Ni^{2+} + 2M^+ + 2SO_4^{2-} + 6H_2O$ ($v = 5$), it is possible to obtain for M^+ ions (b = 2): $D_{2/1} = (m_{01}/m_{02})^{1.25}$ **(Fig. 3)** and for M^{2+} ions (b = 1): $D_{2/1} = (m_{01}/m_{02})^{2.5}$ **(Fig 4)**.

However, for a similar, but simple salt ($NiSO_4 \cdot 7H_2O$), an analogous dependence does not exist (Fig. 5) ($\rho_{xy} = 0.201$ is insignificant).

This is because of the significant differences in the crystal system of proper sulfate hydrates, while all of the investigated double salts are isomorphous, of the same space group ($P2_1/a$) and of almost identical unit cell parameters (a, b, c, β) (their relative standard deviations do not exceed 0.8%)

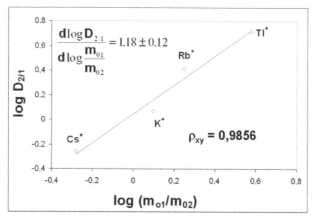

Fig. 3. The dependence of coefficients, $D_{2/1}$ of co-crystalliztion of Cs^+, K^+, Rb^+ and Tl^+ with $NiSO_4 \cdot (NH_4)_2SO_4 \cdot 6H_2O$ on the molality of saturated solutions of suitable salts (Smolik, 1998a)

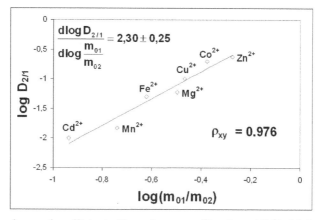

Fig. 4. The dependence of coefficients, $D_{2/1}$ of co-crystallization of Cd^{2+}, Mn^{2+}, Fe^{2+}, Cu^{2+}, Mg^{2+}, Co^{2+} and Zn^{2+} with $NiSO_4 \cdot (NH_4)_2SO_4 \cdot 6H_2O$ on the molality of saturated solutions of suitable salts (Smolik, 2001)

Chemical, Physicochemical and Crystal –
Chemical Aspects of Crystallization from Aqueous Solutions as a Method of Purification

183

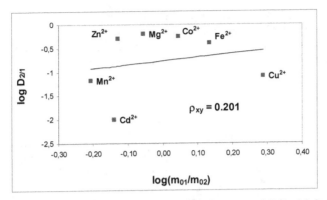

Fig. 5. The dependence of coefficients, $D_{2/1}$ of co-crystallization of Cd^{2+}, Mn^{2+}, Fe^{2+}, Cu^{2+}, Mg^{2+}, Co^{2+} and Zn^{2+} with $NiSO_4 \cdot 7H_2O$ at 20 °C on the molality of saturated solutions of suitable sulfates (Smolik, 2000b]

4.1.2 Crystal system (CS)

The last three examples point to the crystal structure of co-crystallizing salts as a very important factor significantly affecting $D_{2/1}$ coefficients. The dependence of similarity of the crystal structure of macro and microcomponent on mean co-crystallization coefficients in sulfate ($MSO_4 \cdot nH_2O$) crystallization systems is presented in Fig. 6 and Fig. 7.

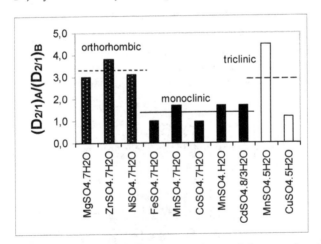

Fig. 6. The dependence of coefficients $D_{2/1}$ on the similarity of the crystal system of macrocomponent (CS_M) and microcomponent (CS_m) ($D_{2/1})_A$ – mean $D_{2/1}$ when (CS_M) = (CS_m) ($D_{2/1})_B$ – mean $D_{2/1}$ when (CS_M) ≠ (CS_m) (Smolik, 2002a, 2004)

As it can be seen, mean $D_{2/1}$ coefficients of microcomponents whose hydrates belong to the same crystal system as the macrocomponent are ~ 3 times (for orthorhombic and triclinic macrocomponents) and ~1.5 times (for monoclinic macrocomponents) greater than those whose hydrates belong to a different crystal system than that of macrocomponent) (Fig. 6).

The mean coefficients $D_{2/1}$ of microcomponents belonging to the same crystal system as the macrocomponent are the highest, and they drop as the similarity of their crystal structure and that of the macrocomponent decreases (taking into account the following direction of the increase of crystal systems symmetry: triclinic<monoclinic<orthorhombic) (Fig. 7).

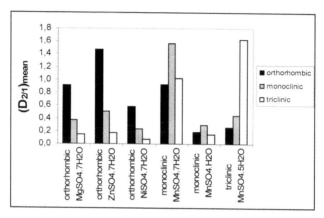

Fig. 7. The dependence of coefficients $D_{2/1}$ on the similarity of crystal systems of macrocomponent (CS_M) and microcomponent (CS_m) ($D_{2/1}$)$_{mean}$ – mean $D_{2/1}$ of microcomponents belonging to the same crystal system (Smolik, 2002a, 2004)

4.1.3 Number of molecules of crystallization water (n)

During the crystallization of hydrates the number of molecules of crystallization water (n) is an additional factor which may influence co-crystallization coefficients $D_{2/1}$. It affects the coordination environment of the metal ion, which in the case of hepta or hexahydrates consists of only water molecules (the linkage of coordination octahedra in these crystals'

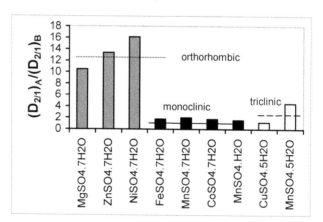

Fig. 8. $D_{2/1} = f(\Delta|n| = |n_M - n_m|)$ $n_M(n_m)$ – number of molecules of crystallization water of macrocomponent (microcomponent) (Smolik, 2002a, 2004),$(D_{2/1})_A$ – mean $D_{2/1}$ when $|\Delta n| = 0$ $(D_{2/1})_B$ – mean $D_{2/1}$ when $|\Delta n| \neq 0$

structures is determined by weak hydrogen bonds), but in lower hydrates oxygen atoms of polyatomic anions enter the coordination environment of the metal ions (causing the formation of chains, closed rings, planar or space networks by vertices–sharing coordination polyhedral) (Balarew, 1987). Generally, mean $D_{2/1}$ values are higher the more similar are n values of macro and microcomponents (Fig. 8 and Fig. 9).

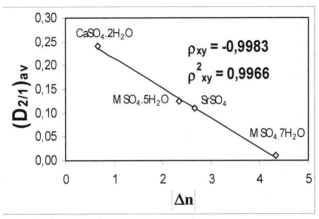

Fig. 9. $(D_{2/1})_{av} = f(\Delta|n| = |n_M - n_m|)$ during the crystallization of $CdSO_4 \cdot 8/3H_2O$ at 20 °C (Smolik, 2002b, 2004), ρ_{xy} – correlation of $(D_{2/1})_{av}$ and $\Delta|n|$

4.1.4 Reciprocal solubility in the solid phase C^s_{MAX}

It is known that the coefficients of co-crystallization of impurities are proportional to their solubilities in the solid phase in the case of crystallization of Ge and Si (Fisher, 1962), as well as several dozen molten metals (Vachobov et al., 1968). However, such regularity has not

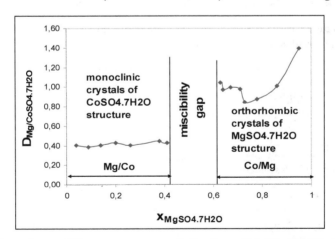

Fig. 10. The dependence of $D_{Mg/CoSO4.7H2O}$ on mole fraction of $MgSO_4 \cdot 7H_2O$ in the solid phase (Oikova et al., 1976) ; Mg/Co – solubility of $MgSO_4 \cdot 7H_2O$ in $CoSO_4 \cdot 7H_2O$, Co/Mg – solubility of $CoSO_4 \cdot 7H_2O$ in $MgSO_4 \cdot 7H_2O$

been found yet during the crystallization of salts from aqueous solutions. The term "solubility in the solid phase" is explained in Fig 10. It is the maximal concentration of a hydrate in another hydrate, which does not cause the change in its structure.

The effect of the solubility in the solid phase (C^s_{MAX}) on $D_{2/1}$ coefficients is presented in Table 5. As it can be seen in most analysed cases, $D_{2/1}$ is proportional to the maximal reciprocal solubility in the solid phase (C^s_{MAX}.) which is proved by relatively high and significant correlation coefficients (ρ_{xy}), marked bold.

No	Macrocomponent	Microcomponents	ρ_{xy}
1.	$MgSO_4 \cdot 7H_2O$	Ni^{2+}, Mn^{2+}, Fe^{2+}, Cu^{2+}, Zn^{2+}, Cd^{2+}, Co^{2+}	**0.7635**
2.	$ZnSO_4 \cdot 7H_2O$	Ni^{2+}, Mn^{2+}, Fe^{2+}, Cu^{2+}, Mg^{2+}, Cd^{2+}, Co^{2+}	**0.8608**
3.	$NiSO_4 \cdot 7H_2O$	Zn^{2+}, Fe^{2+}, Cu^{2+}, Mg^{2+}, Cd^{2+}, Co^{2+}	**0.8455**
4.	$CoSO_4 \cdot 7H_2O$	Ni^{2+}, Mn^{2+}, Fe^{2+}, Mg^{2+}, Cd^{2+}, Zn^{2+}	**0.9172**
5.	$MnSO_4 \cdot 5H_2O$	Zn^{2+}, Cu^{2+}, Mg^{2+}	**0.9971**
6.	$FeSO_4 \cdot 7H_2O$	Ni^{2+}, Zn^{2+}, Cu^{2+}, Mg^{2+}, Cd^{2+}, Co^{2+}	**0.7983**
8.	$Ni(NO_3)_2 \cdot 6H_2O$	Mn^{2+}, Zn^{2+}, Mg^{2+}, Co^{2+}	**0.7022**
9.	$Ni(NO_3)_2 \cdot 6H_2O$	Mn^{2+}, Mg^{2+}, Co^{2+}	**0.9970**
10.	$NiCl_2 \cdot 6H_2O$	Zn^{2+}, Mn^{2+}, Fe^{2+}, Cu^{2+}, Co^{2+}	**0.9974**
11.	K_2SO_4	Cs^+, Tl^+, Rb^+	0.1041

Table 5. Correlation (ρ_{xy}) of co-crystallization coefficients $D_{2/1}$ and solubility in the solid phase (C^s_{MAX}) (Smolik, 2004)

4.1.5 The volume of one formal molecule

The volume of one formal molecule can be calculated, knowing the molar mass of crystallizing salt (compound) and its density, by the following formula $\eta^3 = 10^{24} \dfrac{M_x}{d \cdot N}$ [Å^3] where: M_x – molar mass [g/mole], d – density [g/cm³], N – Avogadro number = $6{,}022 \cdot 10^{23}$/mole). This is very close to that calculated using unit cell parameters (a, b, c, α, β, γ).

Macro-component	Micro-components	y	$\left\| \dfrac{\Delta\eta^3}{\eta^3} \right\|$	$\left(\dfrac{\Delta\eta}{\eta} \right)^2$	Ref.
			x		
			ρ_{xy}		
(1)	Ni^{2+}, Cu^{2+}, Co^{2+}, Zn^{2+}, Mn^{2+}, Cd^{2+}, Ca^{2+} Sr^{2+}	$\ln D_{2/1}$	**-0.8657**		(Smolik, 2008)
	Ni^{2+}, Co^{2+}, Mn^{2+}	$\pi = \ln D_{2/1} -$ $\ln(m_{01}/m_{02})^3$		**-0.9973**	
(2)	Mg^{2+}, Co^{2+}, Ni^{2+}	$\pi = \ln D_{2/1} - \ln$ $(m_{01}/m_{02})^3$		**-0.9987**	(Smolik, 2011)

Table 6. The dependence of $\ln D_{2/1}$ ($\pi = \ln D_{2/1} - \ln (m_{01}/m_{02})^3$) on various functions of η during the crystallization of $Mg(CH_3COO)_2 \cdot 4H_2O$ (1) and $Mn(CH_3COO)_2 \cdot 4H_2O$ (2) at 25 ºC

According to Urusov (1977) this parameter is better than ionic radius in the evaluation of the effect of geometric factor on $D_{2/1}$ coefficients, because it takes into account real interatomic distances defined by crystal system and unit cell parameters. However, η is unambiguous only in ionic crystals of high (cubic) symmetry and in the case of complicated heterodesmic structures of low symmetry these distances become equivocal.

The significant effect of this factor has occurred in some acetate crystallization systems (Table 5).

4.2 Chemical, physicochemical and crystal–chemical properties of co-crystallizing ions

4.2.1 Charge of cation

The cation charge is one of the most important factors influencing $D_{2/1}$ coefficients. Taking into account the earlier mentioned Latin rule "Similia similibus solvuntur" it might be expected that microcomponent ions having the same charge as that of macrocomponent ion should co-crystallize in higher degree than those of different ion charge. In many crystallization systems this rule is fulfilled, e.g., in the case of crystallization of $FeSO_4 \cdot 7H_2O$ at 20 °C the mean $D_{2/1}$ of M^{2+} ions (of the same charge as the macrocomponent Fe^{2+}) are ~14 (50) times greater than those of $M^+(M^{3+})$ ions (of the charge different from that of macrocomponent) (Fig. 11a), and in the case of $Fe(NH_4)(SO_4)_2 \cdot 12H_2O$ crystallization at 20 °C the mean $D_{2/1}$ of M^{2+} ions (of different charge from macrocomponent ions NH_4^+ and Fe^{3+}) is ~400 – 1000 times lower than those of M^+ and M^{3+} ions (Fig 11b).

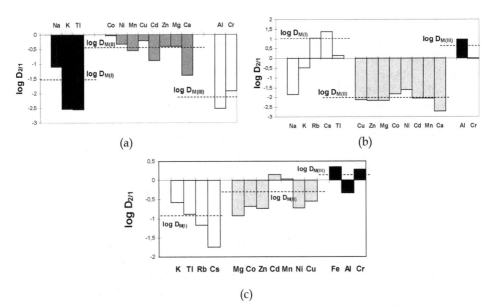

Fig. 11. The effect of ion charge on $D_{2/1}$ during crystallization of: a) $FeSO_4 \cdot 7H_2O$ at 20 °C (Smolik & Lipowska, 1995); b) $Fe(NH_4)$ alum at 20 °C; (Smolik, 1995a); c) Na_2SO_4 at 50 °C; (Smolik, 1998b); log $D_{M(I)(M(II), M(III))}$ – logarithms of mean $D_{2/1}$ for $M^+(M^{2+}, M^{3+})$ ions

However, there are crystallization systems where this simple and evident rule is not fulfilled, e.g., during the crystallization of Na_2SO_4 at 50 °C (Fig. 11c). In this case the mean $D_{2/1}$ of M^+ ions (of the same charge as the macrocomponent Na^+) is the lowest. This is caused by the formation of double, less soluble salts by Na_2SO_4 with M (II) and M (III) sulfates which, because of their structures, are capable of in–build into Na_2SO_4 crystal.

4.2.2 Geometrical factor (ionic radius (r))

The geometrical factor, determined by the difference in size of mutually substituting ions, has been considered for a long time as one of the most significant factors affecting the existence of isomorphism. Beginning from the empirical Goldschmidt rule postulating the border of 15% relative difference of ionic radii for the occurrence of isomorphic substitution, various values of this border (e.g., 5%) have been given by other authors. In addition it has occurred that they are dependent on other factors. This parameter was not recommended by Urusov (1977), who preferred to take into consideration the differences in interionic distances in the solid phase. In this way he calculated $D_{2/1}$ values for M^+ ions were strictly consistent with those experimental ones during the crystallization of alkali metal halides from melt (Urusov & Kravchuk, 1976). However, as proved by the same author, correlation of interionc distances in the solid phase and $D_{2/1}$ values obtained during the crystallization from aqueous solutions occurred as significantly weaker (Urusov, 1980). Moreover, the results of many investigations [(Smolik, 1993, 1995, 1998a, 2003,2007, 2010) indicate that ionic radius can be a useful parameter in the evaluation of co-crystallization coefficients. Some typical dependences of $D_{2/1}$ coefficients on ionic radii have been presented in Fig. 12 a,b,c,d.

We can observe some types of the dependence of $D_{2/1}$ coefficients on ionic radius: a) monotonic, hyperbolic drop of $D_{2/1}$ with the increase of ionic radius (if the case of Cu^{2+} is ignored, because of the structure of triclinic $CuSO_4$ $5H_2O$ significantly departing from the structures of other sulfates); b) the existence of the maximum of $D_{2/1}$ coefficients for ions, whose radii are closest to the radius of macrocomponent and the monotonic drop of $D_{2/1}$ as the absolute value of the difference in ionic radii of macrocomponent and microcomponent increases; c) similar type like "b", disturbed in the case of (mainly) Cu^{2+} because of the almost identical structure of triclinic $CuSO_4$ $5H_2O$ and $MnSO_4$ $5H_2O$; d) there are two ranges of higher $D_{2/1}$ coefficients corresponding to the values of ionic radii very close to those of two macrocomponent ions NH_4^+ and Fe^{3+}.

Very high correlation coefficients of $\ln D_{2/1}$ and $|\Delta r/r_M|$ or $(\Delta r/r_M)^2$ in some crystallization systems (Table 7) indicate that these co-crystallization coefficients strongly depend on the similarity of ionic radii of micro and macrocomponents.

Coefficients of Ca^{2+} co-crystallization with various sulfate hydrates MSO_4 nH_2O, taking into account slight solubility of $CaSO_4$ $2H_2O$, should be very high. However, because of its large radius r_{Ca2+} , Ca^{2+} ion cannot in–build into $MSO_4.nH_2O$ crystals of smaller M^{2+} ions. Therefore, the ionic radius of the macrocomponent is a more important factor than the solubility determining $D_{Ca/MSO4.nH2O}$ values (Fig. 13) (Smolik, 2004).

Thus, ionic radii in the case of crystallization of salts from aqueous solutions can be a convenient and important parameter for the investigation and sometimes evaluation of $D_{2/1}$

Chemical, Physicochemical and Crystal –
Chemical Aspects of Crystallization from Aqueous Solutions as a Method of Purification

189

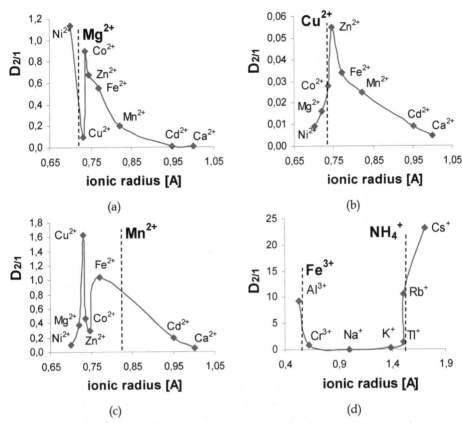

Fig. 12. The effect of ionic radius on coefficients $D_{2/1}$ during the crystallization of : a –
$MgSO_4 \cdot 7H_2O$ at 25 °C (Smolik, 1999a), b – $CuSO_4 \cdot 5H_2O$ at 25 °C (Smolik & Zolotajkin, 1993),
c – $MnSO_4 \cdot 5H_2O$ at 20 °C (Smolik et al., 1995), d – NH_4Fe alum at 20 °C (Smolik, 1995c)

Macro-component	Microcomponents (M^{2+})	Correlation coefficients ρ_{xy} of $\ln D_{2/1}$ and:	
		$\|\Delta r/r_M\|$	$(\Delta r/r_M)^2$
Orthorhombic $MgSO_4 \cdot 7H_2O$	Ni^{2+}, Cu^{2+}, Co^{2+}, Zn^{2+}, Fe^{2+}, Mn^{2+}, Cd^{2+}, Ca^{2+}	**–0.9061***	
	Co^{2+}, Fe^{2+}, Mn^{2+}, Cd^{2+}, Ca^{2+} (monoclinic)	**–0.9992**	
triclinic $MnSO_4 \cdot 5H_2O$	Ni^{2+}, Cu^{2+}, Co^{2+}, Zn^{2+}, Fe^{2+}, Mn^{2+}, Cd^{2+}, Ca^{2+}		–0.8353
	Co^{2+}, Fe^{2+}, Cd^{2+}, Ca^{2+} (monoclinic)		**–0.9948**
triclinic $CuSO_4 \cdot 5H_2O$	Ni^{2+}, Cu^{2+}, Co^{2+}, Zn^{2+}, Fe^{2+}, Mn^{2+}, Cd^{2+}, Ca^{2+}		–0.6659
	Co^{2+}, Fe^{2+}, Mn^{2+}, Cd^{2+}, Ca^{2+} (monoclinic)		**–0.9949**

* –significant ρ_{xy} (for the confidence level of 0.95) are marked bold (Smolik, 2004)

Table 7. Correlation coefficients (ρ_{xy}) of $\ln D_{2/1}$ and $(\Delta r/r_M)^2$ (or $\|\Delta r/r_M\|$) in some sulfate
crystallization systems for all ions or those ions whose sulfate hydrates are monoclinic

coefficients. The comparison of correlation coefficient (ρ_{xy}) of $\pi = [\ln D_{2/1} - \ln(m_{01}/m_{02})^3][y]$ and $|((r_{Co2+})^3 - (r_{M2+})^3)/(r_{Co2+})^3|[x]$ ($\rho_{xy} = -0.9174$) with analogous correlation coefficient of π [y] and $|((\eta_{Co2+})^3 - (\eta_{M2+})^3)/(\eta_{Co2+})^3|[x]$ ($\rho_{xy} = -0.8590$) indicates that the latter parameter, preferred by Urusov (1977) to estimate $D_{2/1}$ values, is in the case of $Co(CH_3COO)_2 \cdot 4H_2O$ crystallization not better than the first one related to ionic radius (Smolik et al., 2007).

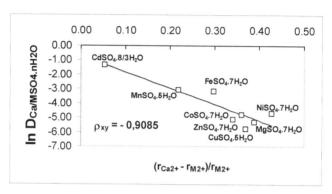

Fig. 13. The dependence of $\ln D_{Ca/MSO4.nH2O}$ on ionic radii of macrocomponent ions (M^{2+}) (Smolik, 2004)

4.2.3 Electronegativity of mutually substituting elements $\Phi(\varepsilon)$

Apart from the geometric factor, the partition coefficients may be affected by the nature (polarity) of the chemical bond of mutually substituting components. The exact quantitative characteristic of the polarity of the chemical bond is given by the integral of overlapping of atomic orbitals, but such data for many isomorphous systems are not available yet. So some authors (Urusov, 1977; Ringwood, 1955) consider the difference in electronegativity of elements as a measure of polarity.

Spectacular examples of a huge effect of this factor on $D_{2/1}$ are given by (Urusov, 1977) where during the crystallization from the melt in systems NaBr – AgBr, NaCl – AgCl, and NaCl – CuCl, co-crystallization does not take place, although the relative differences of interionic distances of co-crystallizing isomorphous salts are very low or close to zero. The difference in the effective ionic charges of Ag(Cu) and Na in these systems was believed to be responsible for the extremely low miscibility of these systems (Kirkova et al., 1996).

In the case of crystallization of several groups of salts from aqueous solutions, the effect of electronegativity of mutually substituting components on distribution coefficients has been compared with the effect of ionic radius. However, for this comparison function $\pi = \ln D_{2/1} - \ln(m_{01}/m_{02})^{v/b}$ has been taken into consideration (it allows for the solubility ratio of co-crystallizing salts). The results are given in Table 8.

As it can be seen, higher correlation coefficients $|\rho_{xy}|$ of π and $\Phi(\varepsilon)$ than those of π and $f(r)$ occur in the case of chlorides ($MCl_2 \cdot nH_2O$) and formates ($M(HCOO)_2 \cdot 2H_2O$). Generally, more significant correlation coefficients $|\rho_{xy}|$ occur in the case when in coordination surroundings of appropriate cations M^{n+} in the solid phase, halogen or formate ions appear as well, apart from oxide ligands originating from water or inorganic anions. However, a kind of dependence of $D_{2/1}$ coefficients on electronegativity has not been univocally defined and the kind of function $\Phi(\varepsilon)$ is very important to decide if unilateral isomorphism occurs, in spite of a buffering action of surrounding hydrate mantle (Kirkova et al., 1996).

Groups of considered salts	maximal values of correlation coefficients $\lvert \rho_{xy} \rvert$ of function $\pi = \ln D_{2/1} - \ln(m_{01}/m_{02})^{v/b}$	
	and $f(r)$	and $\Phi(\varepsilon)$
$MSO_4 \cdot nH_2O$	0.7309	0.6377
$M(NO_3)_2 \cdot nH_2O$	0.9421	0.6502
$MCl_2 \cdot nH_2O$	0.6258	0.7328
$M^I M^{III}(SO_4)_2 \cdot 12H_2O$, M(I)	0.9188	non-significant
$M^I M^{III}(SO_4)_2 \cdot 12H_2O$, M(III)	0.7651	non-significant
$M^I M^{II}(SO_4)_2 \cdot 6H_2O$, M(II)	0.4921	0.3358
$M(HCOO)_2 \cdot 2H_2O$	non-significant	0.7604
$M^I_2 SO_4 \cdot nH_2O$	0.8005	0.6900
MX	0.8749	0.8745

Table 8. Comparison of absolute maximal values of correlation coefficients of functions $\pi = \ln D_{2/1} - \ln(m_{01}/m_{02})^{v/b}$ and $f(r)$ or $\Phi(\varepsilon)$ in considered groups of salts (Smolik, 2004)

To evaluate if unilateral isomorphism occurs during low temperature crystallization from aqueous solutions (according to the rule known for a long time in geology [Urusov, 1970] that lithophilic elements are substituted in the solid phase by chalcophilic and siderophilic elements and not to the contrary) several criteria may be applied (Smolik, 2004). The strongest of them in the case of co-crystallization of more (M_1) and less (M_2) electronegative ions ($\varepsilon_1 > \varepsilon_2$) looks as follows: $\Theta > 3$, where $\Theta = \theta_{1/2}/\theta_{2/1}$; $\theta_{1/2} = (D_{1/2})_{exp.}/(D_{1/2})_{cal}$ and $\theta_{2/1} = (D_{2/1})_{exp.}/(D_{2/1})_{cal}$; $(D_{1/2})_{cal} = (m_{02}/m_{01})^{v/b}$; $(D_{2/1})_{cal} = (m_{01}/m_{02})^{v/b}$

So:

$$\ln\Theta = \ln\theta_{1/2} - \ln\theta_{2/1} = [\ln(D_{1/2})_{exp} - \ln(m_{02}/m_{01})^{v/b}] - [\ln(D_{2/1})_{exp} - \ln(m_{01}/m_{02})^{v/b}] > 1.1.$$

The value $\theta_{1/2}$ ($\theta_{2/1}$) is a measure of the extension (or diminution) of the experimental $(D_{1/2})_{exp}$ $((D_{2/1})_{exp})$ in relation to $(D_{1/2})_{cal}((D_{2/1})_{cal})$, which may be brought about by the effect of electronegativity. $\theta_{1/2}$ and $\theta_{2/1}$ coefficients allowing for the solubility of corresponding salts are independent of the structure and the relative difference in interionic distances of co-crystallizing isomorphous salts. Therefore, they are most adequate to evaluate the occurrence of unilateral isomorphism. The results of such an analysis of all isomorphous salts forming hydrates, for which $D_{2/1}$ ($D_{1/2}$) coefficients have been available, show the lack of any example of unilateral isomorphism (Smolik, 2004)

4.2.4 Electronic configuration

Electronic configuration of M^{n+} ions, as well as crystal field stabilization energy in high spin octahedral complexes [ML_6], may influence the crystal structure of co-crystallizing salts. According to their electronic configuration these ions may be divided into two groups:

closed shell ions having the configuration p^6 (Mg^{2+}, Ca^{2+}, Sr^{2+}, Ba^{2+}) or d^{10} (Zn^{2+}, Cd^{2+}), as well as d^5, but only when they are in the high spin state (Mn^{2+}). The crystal field stabilization energy (CFSE) of such ions is zero;

open shell ions having the configuration d^n ($n \neq 0, 5, 10$), where CFSE $\neq 0$.

In the first case the energy of these ionic coordination compounds due to the metal ions would be independent of the spatial orientation of the metal–ligand bonds. For this reason these metal ions permit variations over wide ranges of structural parameters, mainly the structure defining angles (angular deformations) (Balarew, 1987).

In the second case the CFSE depends on the orientation of metal–ligand bonds. Therefore, there are some preferred structures, for which CFSE has a maximum value, and the change in geometry of coordination polyhedron with respect to these preferred structures is related to CFSE losses (Balarew, 1987). However, the amount of the CFSE is only 5 - 10% of the whole bonding energy in the crystals and other factors mentioned previously (ionic radii, their charge, energy of metal – ligand bonds) determine the structure of predominantly ionic crystals (Balarew, 1987). Hence its effect on $D_{2/1}$ coefficients is rarely observable.

Some examples of the direct influence of electron configuration of ion on $D_{2/1}$ coefficients are presented in Tables **9-10**.

Ion M^{2+}	Electron configuration	$D_{2/1}$
Mg^{2+}	$1s^2 2s^2 p^6$	0.009 ± 0.005
Ca^{2+}	$1s^2 2s^2 p^6 3s^2 p^6$	0.022 ± 0.008
Sr^{2+}	$1s^2 2s^2 p^6 3s^2 p^6 4s^2 p^6$	0.013 ± 0.006
Zn^{2+}	$1s^2 2s^2 p^6 3s^2 p^6 d^{10}$	0.014 ± 0.005
Cd^{2+}	$1s^2 2s^2 p^6 3s^2 p^6 d^{10} 4s^2 p^6 d^{10}$	0.010 ± 0.006
Cu^{2+}	$1s^2 2s^2 p^6 3s^2 p^6 d^9$	0.040 ± 0.007
Mn^{2+}	$1s^2 2s^2 p^6 3s^2 p^6 d^5$	0.40 ± 0.02
Fe^{2+}	$1s^2 2s^2 p^6 3s^2 p^6 d^6$	1.70 ± 0.20
Co^{2+}	$1s^2 2s^2 p^6 3s^2 p^6 d^7$	2.60 ± 0.30

Table 9. The effect the electron configuration on $D_{2/1}$ coefficients during the crystallization of $NiCl_2 \cdot 6H_2O$ at 20 °C (Smolik, 1999b)

As it can be seen, the direct effect of electronic configuration of microcomponent ions on coefficients $D_{2/1}$ is most distinct in the case of chloride, formate and acetate crystallization systems, where mean $(D_{2/1})_{open\ shell}$ coefficients are several times greater than $(D_{2/1})_{closed\ shell}$ ones. This effect is lower in sulfate and nitrate crystallization systems.

The direct effect of the electron configuration of ions depends on the kind of anion of the crystallizing salt. It is slightly perceptible in the case of nitates and sulfates, where, besides water molecules, oxoanions NO_3^- and SO_4^{2-} appear. The valence available between oxygen and metal ion, which is a measure of their anion base strength equals: 0,33 and 0,50, respectively, and is very close to that of water molecules (0,40) (Balarew, 1987). Because of a great excess of water both in the liquid phase and in the solid one (particularly in hepta and hexahydrates), these anions cannot compete with water molecules in the bonding of metal ion. So the environment around both metal cations will be formed mainly by water molecules.

Chemical, Physicochemical and Crystal –
Chemical Aspects of Crystallization from Aqueous Solutions as a Method of Purification

193

Group of salt	Lewis base strength of anion (x) (Brown, 1981)	Mean coefficients ($D_{2/1}$) for microcomponents (ions)		$\kappa = \dfrac{D^o_{mean}}{D^C_{mean}}$ (y)
		closed shell D^c_{mean}	open shell D^o_{mean}	
$MSO_4 \cdot nH_2O$	0.50	0.28	0.50	1.79
$M(NO_3)_2 \cdot nH_2O$	0.33	0.48	0.75	1.56
$MCl_2 \cdot nH_2O$	1.00	0.02	0.37	18.5
$M(HCOO)_2 \cdot 2H_2O$	0.50	1.43	6.48	4.53
$M(CH_3COO)_2 \cdot nH_2O$	0.55	1.26	8.07	6.40
		$\rho_{xy} = 0.9723$		

Table 10. The effect of electron configuration of microcomponents (ions) on their $D_{2/1}$ coefficients in considered groups of simple salts (Smolik, 2004)

The base strength of anions occurring in chloride, acetate and formate systems is generally higher than that for water molecules and equals: 1,00 for Cl^-, 0,55 for CH_3COO^- and 0,50 for $HCOO^-$ (Balarew, 1987)]. Due to this, as well as because of lower excess of water in relation to CH_3COO^- and $HCOO^-$ in the solid phase, these anions can compete with water molecules in coordination surrounding cations of macro and microcomponents. The presence of both kinds of ligands (water molecule and anions Cl^-, CH_3COO^- or $HCOO^-$ differing in size and charge) causes stronger deformation of the octahedral surrounding of these cations as compared with the presence of one ligand. This deformation depends on the electron configuration of the cation of both the macrocomponent and microcomponent. Therefore, this factor may influence the ability of mutual substitution of those octahedra (deformed to a different degree), whose measure is coefficient $D_{2/1}$.

The dependence of coefficients $D_{2/1}$ on electron configuration is usually connected with their dependence on the crystal field stabilization energy (s), which may be expressed quantitatively for high spin octahedral complexes (most of them occurring in the structures of the considered salts) in kJ/mol or in Dq (where Dq – natural theoretical unit for crystal-field splitting energies (Porterfield, 1993). Thus, it is possible to characterize quantitatively this dependence calculating the correlation coefficients of $\ln D_{2/1}$ or $\pi = \ln D_{2/1} - \ln(m_{01}/m_{02})^{v/b}$ and $\Delta s = s_{MACR} - s_{micr}$, $|\Delta s|$, or $(\Delta s)^2$.

Number of crystallization systems	correlation coefficients (ρ_{xy}) of $\pi = \ln D_{2/1} - \ln (m_{01}/m_{02})^3$ and				
	$f(r)$	$f(\delta)$	Δn	$\Delta \varepsilon$	Δs
37	0.4156	−0.1998	−0.1170	**0.7604**	**0.8486**

Table 11. Comparison of correlation coefficients (ρ_{xy}) of $\pi = \ln D_{2/1} - \ln (m_{01}/m_{02})^3$ and functions of some factors affecting co-crystallization coefficients $D_{2/1}$ in formate crystallization systems (Smolik, 2004)

The direct effect of Δs on $\ln D_{2/1}$ is very slight in most considered groups of salts, but having taken into account the solubility ratio of the co-crystallizing salts, the found correlation coefficients of $\pi = \ln D_{2/1} - \ln (m_{01}/m_{02})^{v/b}$ and Δs are relatively high only for formate crystallization systems ($\rho_{xy} = $ **0.8486**). In this group of salts, this correlation coefficient of π and Δs is the highest as compared to the ones involving all analysed factors (Table **11**).

Significant correlation coefficients of $\pi = \ln D_{2/1} - \ln (m_{01}/m_{02})^{v/b}$ and $|\Delta s|$ or $(\Delta s)^2$ occur in none of the considered groups of salts, which means that π does not depend (in them) on the similarity of the CFSE of the macrocomponent and microcomponent ion.

4.2.5 Cation hardness (h)

The concept of cation and anion hardness introduced by Pearson (1963) was utilized by Balarew and co-workers (1984) to solve some crystal–chemical problems. Based on a quantitative definition of hardness (Klopman, 1968) and using his procedure they determined hardness of several open shell cations (Mn^{2+}, Fe^{2+}, Co^{2+}, Ni^{2+} i Cu^{2+}), which together with the values given by Klopman for anions and other cations they used for the anticipation of a kind of coordination polyhedra in these compounds and their structure. The hardness of several other cations was given by Tepavičarova et al. (1995).

The hardness of cations affecting their coordination surrounding in the case of several ligands of different anion hardness may change the crystal structure of appropriate hydrates causing the formation of pure simple salts, solid solutions or double salts (Balarew, 1987) and therefore, influencing $D_{2/1}$ coefficients.

Absolute values of the determined correlation coefficients (ρ_{xy}) of $\ln D_{2/1}$ and h are generally low (Smolik, 2004). However, after taking into account the solubility ratio of the co-crystallizing salts (function $\pi = \ln D_{2/1} - \ln (m_{01}/m_{02})^{v/b}$), ρ_{xy} values are significant for sulfates (MSO_4 nH_2O and M_2SO_4 xH_2O), chlorides (MCl_2 nH_2O) and alkali halides MX (X = Cl^-, Br^-, I^-), particularly in the presence of Tl^+ and become the highest $(|(\rho_{xy})_{sr}| > 0.80)$ and significant in chloride ($MCl_2 \cdot nH_2O$) and halide (MX) crystallization systems. In each case they are negative, which means that with the increasing similarity of macro and microcomponent (with regard to hardness) π values grow.

Low values of ρ_{xy} occur in groups of salts, where coordination surrounding of cations is homogeneous. It is composed of water molecules, which are hard ligands and anions NO_3^-, SO_4^{2-} and CH_3COO^-, classified as hard bases. In chloride crystallization systems ($MCl_2 \cdot nH_2O$) anions Cl^- occur, whose hardness is less than that of water molecules, and in halogen crystallization systems the cation surrounding in the solid phase consists only of chloride, bromide and iodide anions, which are classified as decidedly soft anions. The greatest effect of hardness on co-crystallization coefficients appears here (particularly in the presence of Tl^+ ions significantly differing in their hardness from alkali ions).

5. Possibility of estimation of $D_{2/1}$ coefficients basing on the determined dependences

The determined correlation coefficients of $D_{2/1}$ or $\pi = \ln D_{2/1} - \ln(m_{01}/m_{02})^{v/b}$ and various functions of ionic radii (r), electronegativity (ε), crystal field stabilization energy (s), hardness of cations (h), number of molecules of crystallization water (n), the volume of one formal molecule of hydrate salt (η^3) permit to find a kind of functions (f, φ, ψ, …) of these parameters and to estimate coefficients (w_1, w_2, w_3, …) in the general equation (19) for the evaluation of $D_{2/1}$ coefficients. Some particular equations of such a general type for the estimation of $D_{2/1}$ in several groups of crystallization systems (macrocomponent – microcomponents) at average error not exceeding 31% are presented in Table 12.

Chemical, Physicochemical and Crystal –
Chemical Aspects of Crystallization from Aqueous Solutions as a Method of Purification

195

Macro-component	Micro-components (M^{n+}) (1)	k	Equation	Δ_{av} [%]
orthorhombic $M'SO_4 \cdot 7H_2O$ $M' = \{Mg, Zn, Ni\}$	$Co^{2+}, Fe^{2+}, Mn^{2+}, Cd^{2+}$	21	$D_{2/1} = 0.362 \cdot \left(\dfrac{m_{01}}{m_{02}}\right)^2 \cdot \exp\left(-25.95 \cdot \left(\dfrac{r_1 - r_2}{r_1}\right)^2\right) + 0,0072$	21.6
	Cu^{2+}		$D_{2/1} = 0.020 \cdot \left(\dfrac{m_{01}}{m_{02}}\right)^2 \cdot \exp\left(-25.95 \cdot \left(\dfrac{r_1 - r_2}{r_2}\right)^2\right) + 0,0072$	
	$Ni^{2+}, Mg^{2+}, Zn^{2+}$		$D_{2/1} = \left(\dfrac{m_{01}}{m_{02}}\right)^2 \cdot \exp\left(-25.95 \cdot \left(\dfrac{r_1 - r_2}{r_1}\right)^2\right) + 0,0072$	
orthorhombic $M'(NO_3)_2 \cdot 6H_2O$ $M' = \{Zn, Mn\}$	$Ni^{2+}, Mg^{2+}, Zn^{2+}, Co^{2+}, Mn^{2+}$	8	$D_{2/1} = \left(\dfrac{m_{01}}{m_{02}}\right)^3 \cdot \exp\left(-55.83 \cdot \left(\dfrac{r_1 - r_2}{r_1}\right)^2\right) + 0.143$	14.8
$CoCl_2 \cdot 6H_2O$	$Ni^{2+}, Mg^{2+}, Zn^{2+}, Co^{2+}, Cu^{2+}, Mn^{2+}, Cd^{2+}, Ca^{2+}, Sr^{2+}$	9	$D_{2/1} = \exp\left(0.0078\eta^3 - 1.42|\Delta h| + 0.32\Delta s - 2.92\right)$	27.0
alums	$Al^{3+}, Fe^{3+}, Cr^{3+}$	9	$D_{2/1} = \left(\dfrac{m_{01}}{m_{02}}\right)^2 \cdot \exp\left(16.85 \cdot \eta - 130.2\right)$	29.8
M^I_2 $M^{II}(SO_4)_2 \cdot 6H_2O$ $M^I = \{NH_4^+, Rb^+\}$	$Na^+, K^+, Rb^+, Cs^+, Tl^+,$	7	$D_{2/1} = \left(\dfrac{m_{01}}{m_{02}}\right)^{1.25} \cdot \exp\left(1.48 \cdot \eta - 10.2\right)$	9.1
M^I_2 $M^{II}(SO_4)_2 \cdot 6H_2O$ $M^I = \{Ni, Mg, Cu, Co, Zn, Fe, Mn\}$	$Ni^{2+}, Mg^{2+}, Zn^{2+}, Mn^{2+}, Co^{2+}, Cu^{2+}, Cd^{2+}$	43	$D_{2/1} = \left(\dfrac{m_{01}}{m_{02}}\right)^{2.5} \cdot \exp\left(0.835 \cdot \dfrac{1}{r_2^2} - 0.082\right)$	25.9
$M'(HCOO)_2 \cdot 2H_2O$ $M^I = \{Ni, Mg, Co, Zn, Fe, Mn, Cd\}$	$Ni^{2+}, Mg^{2+}, Zn^{2+}, Mn^{2+}, Co^{2+}, Cu^{2+}, Cd^{2+}$	37	$D_{2/1} = \left(\dfrac{m_{01}}{m_{02}}\right)^3 \cdot \exp\left(1.67\dfrac{\eta_1^3 - \eta_2^3}{\eta_1^3} + 1.37\Delta\varepsilon - 0.057\Delta h + 1.65\Delta s + 0.153\right)$	31.0
$M'_2SO_4 \cdot nH_2O$ $M' = \{Na, K, Tl\}$	$Na^+, K^+, Rb^+, Cs^+, Tl^+,$	9	$D_{2/1} = \exp\left(12.88\dfrac{1}{r_2^2} - 4.829\Delta\varepsilon - 7,684\right)$	22.8 (4)
MX $X = \{Cl, Br, I\}$	Cs^+, Rb^+, K^+	9	$D_{2/1} = \left(\dfrac{m_{01}}{m_{02}}\right)^2 \cdot \exp\left(-132.8\left(\dfrac{r_1 - r_2}{r_1}\right)^2 + 4.38\Delta h + 0.072\right)$	29.3 (4)
MNO_3 $M' = \{K, Rb, Cs\}$	Cs^+, Rb^+, K^+	5	$D_{2/1} = \exp\left(-16.57\dfrac{1}{r_2} - 349.4(\Delta\varepsilon)^2 + 9,209\right)$	18.9

Table 12. Part 1. Some examples of the estimation of coefficients $D_{2/1}$ (Smolik, 2004)

$MClO_3$ $M' =\{K, Rb\}$	Cs^+, Rb^+, K^+	3	$D_{2/1} = \left(\dfrac{m_{01}}{m_{02}}\right)^2 \cdot \exp\left(-4.282 \cdot r_2 + 4.074\right)$	5.3
$MClO_4$ $M' =\{K, Rb, Cs\}$	Cs^+, Rb^+, K^+	5	$D_{2/1} = \left(\dfrac{m_{01}}{m_{02}}\right)^2 \cdot \exp\left(-74.20 \cdot \left(\dfrac{r_1 - r_2}{r_1}\right)^2 + 2.345\Delta h - 0,185\right)$	26.5
M_2CrO_4 $M' =\{K, Rb, Cs\}$	Cs^+, Rb^+, K^+	4	$D_{2/1} = \left(\dfrac{m_{01}}{m_{02}}\right)^2 \cdot \exp\left(-7.967 \cdot \left(\dfrac{r_1 - r_2}{r_1}\right) - 0.424\right)$	5.1

[1] – M^{n+} is not microcomponent, when M'=M;

[2] – subscripts "1" and "2" relate to macrocomponent and microcomponent respectively;

[3] – $\eta = 10^8 \sqrt[3]{\dfrac{M}{D_x N}}$ [Å], where M – molar mass of salt [g/mol], D_x – its density [g/cm³],

N – Avogadro number [$6.022 \cdot 10^{23}$/mol];

[4] – in this case $|(r_1-r_2)/r_1| <0.20$ (for the other cases $D_{2/1} < 0.06$); k – number of crystallizations systems (macrocomponent)$_i$ – (microcomponent)$_{ii}$ in a given group of salt.

Table 12. Part 2. Some examples of the estimation of coefficients $D_{2/1}$ (Smolik, 2004)

6. Methods of lowering $D_{2/1}$ values as the way of increasing purification efficiency of crystallization

From the practical point of view it is very interesting to ascertain how to increase the efficiency of crystallization purification of inorganic substances. This is possible when co-crystallization coefficients $D_{2/1}$ can be lowered. They depend generally, as shown above, on chemical, physicochemical and crystal–chemical properties of co-crystallizing salts and ions. However, there are some previously mentioned "external" factors (such as the kind and composition of the solvent – the liquid phase, the presence of ions or other foreign substances, the presence of complexing agents, acidity [pH] of the solution, from which crystallization takes place, temperature) which may influence these coefficients. Their effect on $D_{2/1}$ coefficients will be discussed below.

6.1 The effect of the kind and composition of the solvent – the liquid phase

The change of composition of the solvent, from which the crystallization takes place, alternates solubilities of co-crystallizing salts, as well as activity coefficients of all components in the liquid phase and indirectly in the solid phase.

Because of decreased water activity, the formed crystal hydrates have a lower number of molecules of crystallization water of different structures, which may influence the similarity of the crystal structure of macro and microcomponents. All these mentioned factors vary in different directions and to a different degree, and therefore, they may finally cause the change of co-crystallization coefficients. Examples are given in Table **13**.

Chemical, Physicochemical and Crystal –
Chemical Aspects of Crystallization from Aqueous Solutions as a Method of Purification

197

Composition of the mother solution [% v/v]						Co-crystallization coefficients $D_{2/1}$				
H_2O	iso-PrOH	CH_3OH	C_2H_5OH	Et_2O	Me_2CO	Mg^{2+}	Co^{2+}	Fe^{2+}	Mn^{2+}	Cu^{2+}
100						0.50± 0.04	0.96± 0.07	0.60± 0.05	0.22± 0.02	0.21± 0.02
50	20	30				0.42± 0.04	0.80± 0.06	0.31± 0.03	0.10± 0.01	0.14± 0.01
60			38	2		0.39± 0.03	0.96± 0.07	0.46± 0.05	0.14± 0.01	0.17± 0.02
63					37	0.57± 0.05	0.70± 0.06	0.49± 0.04	0.18± 0.02	0.15± 0.02

Table 13. The effect of addition of various organic solvents: iso-propyl alcohol (iso-PrOH), methanol (CH_3OH), ethanol (C_2H_5OH), diethyl ether (Et_2O), acetone (Me_2CO) on coefficients $D_{2/1}$ during the crystallization of $NiSO_4 \cdot 7H_2O$ at 25 ºC (Smolik, 1984)

6.2 The effect of the presence of other ions or substances in the liquid phase

Interactions which happen in the aqueous phase may also influence $D_{2/1}$ coefficients. This effect is formally taken into consideration in equation (19) by the mean activity coefficients of both the macrocomponent (γ_{m1}) and microcomponent (γ_{m2}).

If the action of various factors in aqueous solution causes the same changes in both mean activity coefficients, so that $\frac{\gamma_{m1}}{\gamma_{m2}} = const.$, then at unchanged properties of the solid phase $D_{2/1}$ coefficient remains constant (e.g., the addition of HBr during the co-crystallization of Ra^{2+} with $BaBr_2$ does not affect $D_{2/1}$ coefficient, likewise the introduction of weak electrolytes (glucose or CH_3COONa + CH_3COOH) having no common ions with micro and macrocomponent ($Ra(NO_3)_2$ i $Ba(NO_3)_2$) and not reacting with them also does not change the $D_{2/1}$ value (Chlopin, 1938).

However, if substances are present in the solution that react in a different way with the macro and microcomponent ions forming slightly dissociated compounds, an essential change of $D_{2/1}$ coefficients takes place (e.g., the addition of CH_3COONa + CH_3COOH in the crystallization system $Pb(NO_3)_2$ – $Ra(NO_3)_2$ – H_2O causes bonding a part of Pb^{2+} ions in slightly dissociated acetate, which leads to the lowering of its mean activity coefficient and finally to the increase of radium co-crystallization coefficient (Chlopin, 1938).

6.3 The effect of the presence of complexing agents

The presence of complexing agents has a significant influence on coefficient, $D_{2/1}$. In the case of forming complexes by both macrocomponent and microcomponent, the relationship between the value of co-crystallization coefficient in the presence of complexing agent $(D_{2/1})^k$ and that in the case of its absence $(D_{2/1})$ is expressed by equation (Mikheev et al.,

1962): $D^k_{2/1} = D_{2/1} \cdot \dfrac{1 + \dfrac{[ML]}{[M]}}{1 + \dfrac{\beta}{\beta'} \cdot \dfrac{[ML]}{[M]}}$, where β (β') – stability constant of the complex of microcomponent (macrocomponent), [ML] and [M] are the concentrations of macrocomponent complexes and its free ions. Hence if $\beta/\beta'>1$ (stability of the complex with microcomponent is higher), then $D^k_{2/1} < D_{2/1}$ (microcomponent co-crystallizes to a lower degree and vice versa.

An example of the use of a complexing agent to significantly lower $D_{2/1}$ coefficients is presented in Table 14.

The presence of $EDTA^{4-}$ in stoichiometric amount causes 4 – 180 fold lowering of $D_{2/1}$ coefficients of M^{2+} and M^{3+} ions, because of the formation of $[M(EDTA)]^{2-}$ or $[M(EDTA)]^-$ anionic complexes, and in the case of some of them (Cd^{2+}, Mn^{2+}, Ni^{2+}, Al^{3+} and Cr^{3+}) two-fold increase of $EDTA^{4-}$ excess leads to an additional drop of $D_{2/1}$ coefficients.

Ions	Co-crystallization coefficients , $D_{2/1}$		
	Ratio of the number of moles of $EDTA^{4-}$ to the sum of the number of moles of M^{2+} and M^{3+} ions before crystallization		
	0:1	1:1	2:1
Fe^{3+}	2.24	0.02	0.02
Co^{2+}	0.21	0.01	0.01
Zn^{2+}	0.18	<0.01	<0.01
Cd^{2+}	1.40	0.10	<0.01
Mn^{2+}	1.08	0.27	0.07
Ni^{2+}	0.19	0.01	0.01
Cu^{2+}	0.28	0.01	0.01
Al^{3+}	0.46	0.05	0.01
Cr^{3+}	1.92	0.16	0.03

Table 14. The effect of $EDTA^{4-}$ addition on $D_{2/1}$ coefficients of co-crystallization of M^{2+} i M^{3+} ions with Na_2SO_4 at 50ºC. (1998b)

6.4 The effect of the acidity (pH) of the solution, from which crystallization takes place

Cations of macrocomponent ($[M(H_2O)_x]^{n+}$) and microcomponent ($[M'(H_2O)_{x'}]^{n+}$) present in the solution, from which crystallization usually takes place, may hydrolyze according to the following equations: $[M(H_2O)_x]^{n+} + H_2O \leftrightarrows [M(H_2O)_{x-1}(OH)]^{(n-1)+} + H_3O^+$ and $[M'(H_2O)_{x'}]^{n+} + H_2O \leftrightarrows [M'(H_2O)_{x'-1}(OH)]^{(n-1)+} + H_3O^+$. The degree of hydrolysis depends on $[M(H_2O)_x]^{n+}$ ($[M'(H_2O)_{x'}]^{n+}$) cation acidic strength, meant as Brönstedt acid

($K_{hi} = \{[M(H_2O)_{x-1}(OH)]^{(n-1)+}\} \cdot \{H_3O^+\}/ \{[M(H_2O)_x]^{n+}\}$. If the difference in K_h of both ions is significant (e.g., $K_{h2} >> K_{h1}$), ions ($[M'(H_2O)_{x'-1}(OH)]^{(n-1)+}$) of different charge than those of macrocomponent ($[M(H_2O)_x]^{n+}$) are present in the solution, which in–build into crystals of macrocomponent to a lower degree (e.g., $D_{Fe(III)/NH4Al\ alum} = 0.038 \pm 0.005$ in 0,1 M H_2SO_4 solution and $D_{Fe(III)/NH4Al\ alum} = 0.063\pm 0.009$ in 1,0 M H_2SO_4 solution) (Smolik, 1995b).

In the case of $NiCl_2$ ·$6H_2O$ crystallization with increasing HCl concentration, the lowering of co-crystallization coefficients $D_{2/1}$ of Co^{2+}, Mn^{2+}, Cu^{2+} and Fe^{2+} (Table 15) is caused not only by the rise of acidity of the solution, but also by the formation of chloride complexes at higher Cl^- concentrations. A significant decrease of Mn^{2+} coefficient (D_{Mn}) occurs even at 0.5 M HCl, but that of Fe^{2+} and Co^{2+} only at 5M HCl.

Chemical, Physicochemical and Crystal –
Chemical Aspects of Crystallization from Aqueous Solutions as a Method of Purification

199

Microcomponent	Average $D_{2/1}$ coefficients for HCl concentrations [mol/L]		
	0	0.5	5
Co^{2+}	2.60±0.30	2.30±0.30	1.80±0.10
Mn^{2+}	0.46±0.02	0.20±0.02	0.21±0.01
Cu^{2+}	0.04±0.01	0.04±0.01	0.02±0.01
Fe^{2+}	1.70±0.20	1.20±0.20	0.40±0.08

Table 15. The effect of HCl concentration on $D_{2/1}$ coefficients of some M^{2+} ions during the crystallization of $NiCl_2 \cdot 6H_2O$ at 25 °C (Smolik, 1999b)

6.5 The effect of the change of the oxidation state

In some cases it is possible to change easily the oxidation state of the microcomponent or macrocomponent during or before crystallization. Usually this is accompanied by a significant alteration of $D_{2/1}$ coefficients, which may be utilized for the rise of purification efficiency. Several examples of the change of microcomponent oxidation state are presented in Table 16.

Crystallized salt temperature	Oxidation state of		Factor changing oxidation state	$D_{2/1}$	Ref.
	Macro–component	Micro–component			
$MnSO_4 \cdot 5H_2O$ at 20 °C	Mn^{2+}	Fe^{2+}	H_2O_2	1.04	(Smolik et al., 1995)
		Fe^{3+}		< 0.03	
$CoSO_4 \cdot 7H_2O$ at 20 °C	Co^{2+}	Fe^{2+}	H_2O_2	1.20	(Smolik, 2003)
		Fe^{3+}		< 0.03	
$NH_4Al(SO_4)_2 \cdot 12H_2O$ at 25 °C	Al^{3+}	Fe^{3+}	$NH_2OH.H_2SO_4$	0.04± 0.01	(Smolik, 1995b)
		Fe^{2+}		< 0.01	

Table 16. The effect of the change of oxidation state of microcomponent on $D_{2/1}$ coefficients

The change of the oxidation state of macrocomponent can be used for the purification of iron salts: crystallization of $FeSO_4 \cdot 7H_2O$ at 20 °C permits, with great efficiency, removal of all M^+ and M^{3+} ions (Fig. 12a) and the remaining M^{2+} ions can be easily removed after transferring $FeSO_4 \cdot 7H_2O$ into $NH_4Fe(SO_4)_2 \cdot 12H_2O$ by oxidation and its crystallization at 20 °C (Fig. 12b).

6.6 The effect of temperature

The effect of temperature on $D_{2/1}$ coefficients is very complex. As temperature increases, the solubilities of the macro and microcomponent (m_{01}, m_{02}), the mean activity coefficients in their binary saturated solutions (γ_{m01}, γ_{m02}), the mean activity coefficients in the ternary solution being in equilibrium with their mixed crystal (γ_{m1}, γ_{m2}) (temperature affects dehydration of ions, processes of hydrolysis or complex formation in solution (Kirkova et al., 1996), as well as activity coefficients of both in their solid solution (f_1, f_2) (temperature influences enthalpy of mixing in the solid phase) change in different directions in various crystallization systems. Therefore, both the increase and drop of co-crystallization coefficient may be observed or sometimes the maintenance of its constant value (in the case of compensation of the all mentioned changes). However, the alteration of $D_{2/1}$ runs generally in a continuous manner as long as there are no phase transitions. If a phase transition in the system takes place, a jump change of $D_{2/1}$ appears, connected with the transition from isomorphous co-crystallization into the isodimorphous one. The determination of temperatures at which such a jump change of co-crystallization coefficients takes place, permits finding the temperatures of the phase transitions for many hydrate sulfates (Purkayastha & Das, 1972, 1975), as well as predicting the existence (at specific conditions) of hydrates of some salts, as yet unknown (Purkayastha & Das, 1971).

Some examples of the changes of co-crystallization coefficients $D_{2/1}$ at various temperatures have been presented in Table 17.

(macro-component)	ion	Tempe-rature [°C]	Kind of hydrate (crystal structure) of		$D_{2/1}$	Ref.
			Macro-component	Micro-component		
MnSO$_4$	Cu^{2+}	20	MnSO$_4$ ·5H$_2$O (tcl.)	CuSeO$_4$ 5H$_2$O (tcl.)	1.63	(Smolik, 2004)
		50	MnSO$_4$ ·H$_2$O (mcl.)	CuSeO$_4$ 5H$_2$O (tcl.)	0.15	
ZnSeO$_4$	Cu^{2+}	25	ZnSeO$_4$ 6H$_2$O (tetr.)	CuSeO$_4$ 5H$_2$O (tcl.)	0.51	(Smolik & Kowalik, 2010, 2011)
		40	ZnSeO$_4$ 5H$_2$O (tcl.)	CuSeO$_4$ 5H$_2$O (tcl.)	1.77	
		50	ZnSeO$_4$ ·H$_2$O (mcl.)	CuSeO$_4$ 5H$_2$O (tcl.)	0.12	
	Ni^{2+}	25	ZnSeO$_4$ 6H$_2$O (tetr.)	NiSeO$_4$ 6H$_2$O (tetr.)	2.93	
		40	ZnSeO$_4$ 5H$_2$O (tcl.)	NiSeO$_4$ 6H$_2$O (tetr.)	0.21	
Sr(NO$_3$)$_2$	Pb^{2+}	29	Sr(NO$_3$)$_2$ 4H$_2$O(mcl.)	Pb(NO$_3$)$_2$ (cub.)	0.66	(Niesmie-anov, 1975)
		34	Sr(NO$_3$)$_2$ (cub.)	Pb(NO$_3$)$_2$ (cub.)	3.30	
Na$_2$SO$_4$	Fe^{3+}	25	Na$_2$SO$_4$ (rhomb.)		0.01	(Smolik, 1998b)
		50	Na$_2$SO$_4$ ·10H$_2$O(mcl.)		2.24	

Table 17. The effect of temperature on $D_{2/1}$ coefficients

The observation of the alterations of $D_{2/1}$ coefficients with changing temperature sometimes permits finding such ranges of this parameter, where they are low enough that crystallization purification of the macrocomponent from a given microcomponent will be very effective (Purkayastha & Das, 1972; Smolik & Kowalik, 2010, 2011). In such a manner it turned out to be possible to accomplish essential purification of CoSeO$_4$ from almost all M^{2+} ions (most difficult to remove) solely by the crystallization method (Kowalik et al., 2011).

7. Conclusions

Crystallization of substances from solutions seems still to be a convenient method of their purification, particularly in obtaining of high purity inorganic compounds. The effectiveness of this process depends on the kind of both macrocomponent (1) and microcomponent (2) and can be evaluated by means of co-crystallization coefficient, $D_{2/1}$ (Henderson – Kraček, Chlopin). These coefficients are affected by conditions applied in the crystallization process, but those which are equilibrium ones, depend exclusively on "internal" and "external" factors.

"Internal" factors (resulting from chemical, physicochemical and crystal-chemical properties of co-crystallizing salts and ions to a significant degree) determine the level of $D_{2/1}$

Chemical, Physicochemical and Crystal –
Chemical Aspects of Crystallization from Aqueous Solutions as a Method of Purification

201

coefficients. The investigation involving the effect of these factors is of great importance because it makes it possible to evaluate the usefulness of crystallization in the separation of a given pair of macrocomponent (1) and microcomponent (2).

By means of "external" factors a significant lowering of $D_{2/1}$coefficients is sometimes possible, and thus, the improvement of the efficiency of crystallization purification.

Growing knowledge concerning coefficients $D_{2/1}$ in new crystallization systems, as well as better understanding of the dependences of these coefficients on different factors, permits evaluating in a progressively better way the possibilities of the crystallization method in new crystallization systems and more effective control with "external" conditions to achieve higher yields of crystallization purification, enrichment of trace amounts of rare, scattered elements for preparative or analytical purposes. In addition it also helps in improving the growing of single crystals of specific properties or explaining the genesis of some minerals.

8. References

Balarew, Chr.; Duhlev R. & Spassov D. (1984). Hydrated metal halide structures and the HSAB concept, *Crystal Res. Technol.* Vol. 19(11), pp. 1469

Balarew, Chr. (1987). Mixed crystals and double salts between metal (II) salt hydrates, *Zeitschrift fuer Kristallographie.*, Vol. 181, pp. 35-82

Blamforth, A. W. (1965). *Industrial Crystallization*, Leonard Hill, London 1965, (Great Britain) England

Borneman-Starinkevich, I. D. (1975) Calculation of the crystallochemical formula as one of the methods for the investigation of minerals, In *Izomorfizm Minerallov* (Eds. F. V. Chukhrov, B. E. Borutsky & N. N. Mozgova), pp. 125-33 . Nauka, Moskva

Brown, I. D. (1981). Bond-valence method: an empirical approach to chemical structure and bonding. In *Structure and bonding in crystals*. (Eds. M. O'Keeffe, A. Nawrotsky) Vol. II, p. 1 – 30. New York: Academic Press 1981

Byrappa, K.; Srikantaswamy, S.; Gopalakrishna G. S. & Venkatachalapaty V. (1986) Influence of admixtures on the crystallization and morphology of AlPO$_4$ crystals. *J. Mater.Sci.*, Vol. 21, pp. 2202 - 2206

Chlopin, V. G. (1938). The distribution of electrolytes between solid crystals and liquid phase, *Trudy Rad. Inst. AN SSSR* Vol. 4, pp. 34-79

Chlopin, V. G. (1957). Izbrannye Trudy, (Selected Work) Vol. 6, pp. 173, *Izd. AN SSSR*, Moskva, Russia

Demirskaya, O. V.; Kislomed A.N.; Velikhov Y.N.; Glushova L. V. & Vlasova D. I. (1989) Distribution of impurities during crystallization of potassium dihydrophosphate from aqueous solutions at 25 ºC. *Vysokochistye Veshchestva*, Vol. 1, pp. 14-16

Doerner, H. A. & Hoskins, W. M. (1925). Coprecipitation of radium and barium sulfates, *J. Am. Chem. Soc.*,Vol.47, pp 662-75

Fisher, S. (1962). Correlation between maximum solid solubility and distribution coefficient for impurities in Ge and Si, *J. Appl. Phys.*, Vol.33, pp 1615

Gorshtein, G. I. (1969). In *Methods of obtaining of high-purity inorganic substances* (in Russ.), Izd. Chimia, Sanct Petersburg, Russia, (63-125)

Hill, A. E.; Durham, G. S. & Ricci, J .E. (1940). Distribution of isomorphous salts in solubility equilibrium between liquid and solid phases, *J. Am. Chem. Soc.*, Vol. 62, pp 2723-32

Kirkova, E. (1994). *Vysokochisti Vechtestva – Metodi na poluchavane*, Sofia University Press, Sofia

Kirkova, E.; Djarova, M. & Donkova, B. (1996). Inclusion of isomorphous impurities during crystallization from solutions, *Prog. Crystal Growth and Charact.*, Vol.32, pp 111 -134

Klopman, G. (1968). Chemical reactivity and the concept of charge- and frontier-controlled reaction, *J. Am.Chem.Soc.* Vol.90, pp. 223-234

Kowalik, A.; Smolik M. & Mączka K. (2011). The influence of temperature on the values of distribution coefficients $D_{2/1}$ during the crystallization of $CoSeO_4 \cdot nH_2O$, The 1st international conference on methods and materials for separation processes – *Separation Science – Theory and Practice*, Kudowa Zdrój 2011, Poland

Matusievich, L. M. (1961). Influence of diffusion on the process of joint crystallization of isomorphous salts, *Zh. Neorg. Khim.*, Vol.6, pp. 1020-7

Mikheev, N. B.; Mikheeva, L. M.; Malinin, A. B. & Nikonov, M. D, (1962). Effect of complex formation on the separation of elements by cocrystallization process obeying logarithmic rule, *Zh. Neorg. Khim*, Vol. 7, pp. 2267-2270

Niesmiejanov, A. N. (1975). *Radiochemia,* PWN, Warszawa, Poland

Oikova, T.; Balarew, Chr. & Makarov, L. L. (1976). Thermodynamic study of magnesium sulfate-cobalt sulfate-water and magnesium sulfate-zinc sulfate-water systems at 25 °C, *Zh. Fiz. Khim.*, Vol.50(2), pp. 347-52

Pearson, R . G. (1963). Hard and soft acids and bases, *J. Am. Chem. Soc.* 85(22), 3533-9

Porterfield, W. W. (1993). *Inorganic Chemistry. Unified Approach*, Academic Press, Inc. San Diego, New York, Boston, London, Sydney, Tokyo, Toronto,

Przytycka, R. (1968). Coprecipitation of radioactive elements, (in Polish) *Wiadomości Chemiczne,* Vol. 22, pp. 121-141

Purkayastha, B. C. & Das, N. R. (1971). Transition temperature of heptahydrated copper sulfate *J. Indian Chem. Soc.*, Vol. 48, pp. 70-4

Purkayastha, B. C. & Das, N. R. (1972). Study of transition temperature through mixed crystal formation II. *J. Indian Chem. Soc.*, Vol. 49, pp. 245-50

Purkayastha, B. C. & Das, N. R. (1975). Transition temperature through mixed crystal formation. IV, *J. Radioanal. Chem*, Vol. 25, pp. 35-46

Ratner, A. P. (1933). Theory of the distribution of electrolytes between a solid crystalline and a liquid phase, *J. Chem. Phys.*, Vol.1, pp 789-94

Ringwood, A. E. (1955). Principles governing trace-element distribution during magmatic crystallization. I. Influence of electronegativity. *Geochim. Cosmochim. Acta*, Vol. 7, pp. 139 – 202

Rojkowski, Z. & Synowiec, J.(1991). *Krystalizacja i krystalizatory,* (Crystallization and crystallizers) (in Polish) WNT, ISBN 83-204-1375-3, Warszawa, Poland

Ruff, O.; Ebertt F.& Luft, F. (1928). Röntgenographic methods for determining substances adsorbed on carbon, *Z. anorg. allgem. Chem.*,Vol. 170, pp. 49-61

Smolik, M. (1984). Investigations on the method of preparation of high-purity $NiSO_4 \cdot 7H_2O$, *Internal report of the Department of Inorganic Chemistry and Technology, Silesian University of Technology, Gliwice* unpublished, (in Polish)

Smolik, M. & Zołotajkin, M. (1993). Partition of trace amounts of impurities during the crystallization of $CuSO_4 \cdot 5H_2O$, *Polish J. Chem.* Vol. 67(3), pp. 383-389

Smolik, M, & Lipowska, B. (1995). Partition of trace amounts of impurities during the crystallization of $FeSO_4 \cdot 7H_2O$ at 20°C., *Indian J. Chem.*, Vol. 34A, pp.230-4

Smolik, M.; Zołotajkin, M. & Kluczka, J. (1995). Distribution of trace amounts of impurities during manganese (II) sulfate crystallization at 20° and 2°C., *Polish J. Chem.*, Vol.69, pp.1322-7

Smolik, M. (1995a). Cocrystallization of trace amounts of metal ions with $NH_4Fe(SO_4)_2.12H_2O$, (in Polish) – 5th Symposium *"Industrial crystallization"* Rudy, Institute of Inorganic chemistry, Gliwice, pp. XVIII (1-8), Poland

Smolik, M. (1995b). Partition of trace amounts of impurities during the crystallization of $NH_4Al(SO_4)_2 \cdot 12H_2O$, (in Polish), Scientific Meeting of the Polish Chemical Society, Lublin, Poland

Smolik, M. (1998a). Distribution of cocrystallized microamounts of some M^+ ions during $NiSO_4.(NH_4)_2SO_4.6H_2O$ crystallization., *Acta Chem. Scand.*, Vol.52, pp 891-6

Smolik, M. (1998b). Cocrystallization of trace amounts of metal ions with Na_2SO_4 (in Polish), 6th Symposium *"Industrial crystallization"* Rudy 1998 (*in Polish*), Institute of Inorganic chemistry, Gliwice, pp. 91-101

Smolik, M. (1999a). Partition of microamounts of some M^{2+} ions during $MgSO_4 \cdot 7H_2O$ crystallization. *Austr. J. Chem.* Vol. 52, pp. 425-430

Smolik M. (1999b) Distribution of trace amounts of some M^{2+} ions during nickel chloride crystallization coefficients, *Polish J. Chem.* Vol.73, pp. 2027-2033

Smolik, M.; (2000a). Distribution of microamounts of some M^{2+} ions during $ZnSO_4 \cdot 7H_2O$ crystallization., *Canadian J. Chem.*, Vol.78(7), pp. 993-1002

Smolik, M. (2000b). Distribution of trace amounts of M^{2+} ions during crystallization of $NiSO_4 \cdot 7H_2O$, *Polish J. Chem.*, Vol. 74, pp. 1447-1461

Smolik, M. (2001). Distribution of microamounts of M^{2+} during $NiSO_4 \cdot (NH_4)_2SO_4 \cdot 6H_2O$ crystallization, *Separation Science and Technology* Vol. 36(13), pp. 2959-2969

Smolik, M. (2002a). Factors influencing cocrystallization coefficients D of trace amounts of M^{2+} ions in some sulfate crystallization systems ($M'SO_4$ nH_2O – H_2O), 10th *International Symposium on Solubility Phenomena*, Varna 2002, Abstract Book, pp.37

Smolik, M. (2002b). Cocrystallization of trace amounts of M^{2+} ions with $CdSO_4$ $8/3H_2O$, 10th *International Symposium on Solubility Phenomena*, Varna 2002, Abstract Book, pp. 93

Smolik, M. (2003). Cocrystallization of low amounts of M^{2+} ions during $CoSO_4 \cdot 7H_2O$ crystallization, *J. Chilean Chem. Soc.*, Vol. 48(3), pp. 13-18

Smolik, M. (2004) (Effect of some chemical, physicochemical and crystal-chemical factors on cocrystallization coefficients $D_{2/1}$ of trace amounts of metal ions during the crystallization of chosen salts from water solutions (in Polish), *Zesz. Nauk. Pol. Sl. Chemia*, No 1617, Vol. 148, pp 1-186

Smolik, M. Jakóbik, A. & Trojanowska J. (2007). Distribution of trace amounts of M^{2+} ions during $Co(CH_3COO)_2 \bullet 4H_2O$ crystallization, *Sep. Pur. Techn.*, Vol. 54, pp. 272-276

Smolik, M. (2008). Effect of chemical, physiochemical and crystal-chemical factors on co-crystallization coefficients of trace amounts of metal ions M2+ during the crystallization of selected acetates M(CH3COO)2 nH2O, *Chemik*, Vol.10, pp 526-529

Smolik, M. & Kowalik, A. (2010). Co-crystallization of trace amounts of M^{2+} ions with $ZnSeO_4$ $6H_2O$ at 25°C, *J. Cryst. Growth*, Vol. 312, pp, 611- 616

Smolik, M. & Kowalik, A. (2011a). Equilibrium coefficients of co-crystallization of M2+ ions with $MgSeO4$ $6H2O$ and their dependences on various physicochemical factors, *Crystal Research and Technology*, Vol. 46, pp. 74-9

Smolik, M. & Kowalik, A. (2011b). Coefficients, $D_{2/1}$ of cocrystallization of M^{2+} ions with $ZnSeO_4 \cdot 5H_2O$ at 40 °C and $ZnSeO_4 \cdot H_2O$ at 50 °C and their dependences on various physicochemical and crystal-chemical factors, *J. Cryst. Growth*, Vol. 337, pp. 46-51

Smolik, M. & Siepietowski, L. (2011). Some factors influencing distribution coefficients of trace amounts of M^{2+} ions during the crystallization of $Mn(CH_3COO)_2 \cdot 4H_2O$, *S. Afr. J. Chem.*, Vol. 64, pp. 34-37

Tepavicharova, S.; Balarew, Chr. & Trendafilova S. (1995). Double salts obtained from $Me+X-CuX_2-H_2O$ systems ($Me+ = K^+$, $NH4^+$, Rb^+, Cs^+; $X- = Cl^-$, Br^-), *J. Solid State Chem.* 114, 385-91

Urusov, V. S. (1970). Energy theory of isovalent isomorphism, *Geochimia*, Vol. 4, pp. 510-24

Urusov, V. S. & Kravchuk I. F. (1976). Energy analysis and calculations of partition coefficients of isovalent isomorphous admixtures during crystallization of melts, *Geochimia*, Vol. 8, pp. 1204-23

Urusov, V. S. (1977). *Theory of the Isomorphous Miscibility*, Nauka, Moskva, Russia

Urusov, V. S. (1980). Energy formulation of the problem of equilibrium cocrystallization from a water solution, *Geochimia*, Vol.54, pp. 627-44

Vachobov, A. V.; Khudaiberdiev, V. G. & Vigdorovich, V. N. (1968). Relation between the distribution coefficient for crystallization purification and the limited solubility of the impurity in the solid state, *Dokl. Tadzh. Akad. Nauk*, Vol.11(8), pp 19-22

Zhelnin, B. I. & Gorshtein G. I. (1971). Investigations on equilibrium in the system $MnSO_4$-$MgSO_4$-H_2O at 25 and 100 °C, *Zh. Neorg. Khim.* Vol. 16, pp. 3146-50

Zolotov, Yu. A. & Kuz'min, N. M. (1990). *Preconcentration of trace elements*, Elsevier, Amsterdam, Oxford, New York, Tokyo

Permissions

The contributors of this book come from diverse backgrounds, making this book a truly international effort. This book will bring forth new frontiers with its revolutionizing research information and detailed analysis of the nascent developments around the world.

We would like to thank Dr. Marcello Rubens Barsi Andreeta, for lending his expertise to make the book truly unique. He has played a crucial role in the development of this book. Without his invaluable contribution this book wouldn't have been possible. He has made vital efforts to compile up to date information on the varied aspects of this subject to make this book a valuable addition to the collection of many professionals and students.

This book was conceptualized with the vision of imparting up-to-date information and advanced data in this field. To ensure the same, a matchless editorial board was set up. Every individual on the board went through rigorous rounds of assessment to prove their worth. After which they invested a large part of their time researching and compiling the most relevant data for our readers. Conferences and sessions were held from time to time between the editorial board and the contributing authors to present the data in the most comprehensible form. The editorial team has worked tirelessly to provide valuable and valid information to help people across the globe.

Every chapter published in this book has been scrutinized by our experts. Their significance has been extensively debated. The topics covered herein carry significant findings which will fuel the growth of the discipline. They may even be implemented as practical applications or may be referred to as a beginning point for another development. Chapters in this book were first published by InTech; hereby published with permission under the Creative Commons Attribution License or equivalent.

The editorial board has been involved in producing this book since its inception. They have spent rigorous hours researching and exploring the diverse topics which have resulted in the successful publishing of this book. They have passed on their knowledge of decades through this book. To expedite this challenging task, the publisher supported the team at every step. A small team of assistant editors was also appointed to further simplify the editing procedure and attain best results for the readers.

Our editorial team has been hand-picked from every corner of the world. Their multi-ethnicity adds dynamic inputs to the discussions which result in innovative outcomes. These outcomes are then further discussed with the researchers and contributors who give their valuable feedback and opinion regarding the same. The feedback is then collaborated with the researches and they are edited in a comprehensive manner to aid the understanding of the subject.

Apart from the editorial board, the designing team has also invested a significant amount of their time in understanding the subject and creating the most relevant covers. They scrutinized every image to scout for the most suitable representation of the subject and create an appropriate cover for the book.

The publishing team has been involved in this book since its early stages. They were actively engaged in every process, be it collecting the data, connecting with the contributors or procuring relevant information. The team has been an ardent support to the editorial, designing and production team. Their endless efforts to recruit the best for this project, has resulted in the accomplishment of this book. They are a veteran in the field of academics and their pool of knowledge is as vast as their experience in printing. Their expertise and guidance has proved useful at every step. Their uncompromising quality standards have made this book an exceptional effort. Their encouragement from time to time has been an inspiration for everyone.

The publisher and the editorial board hope that this book will prove to be a valuable piece of knowledge for researchers, students, practitioners and scholars across the globe.

List of Contributors

Elena A. Chechetkina
Institute of General and Inorganic Chemistry of Russian Academy of Sciences, Moscow, Russia

Frantisek Kavicka, Karel Stransky, Bohumil Sekanina, Josef Stetina and Jaromir Heger
Brno University of Technology, Czech Republic

Jana Dobrovska
VSB TU Ostrava, Czech Republic

Abhay Kumar Singh
Department of Physics, Banaras Hindu University, Varanasi, India

Nicole Stieger and Wilna Liebenberg
North-West University, Unit for Drug Research & Development, South Africa

Loredana Tanasie and Stefan Balint
West University of Timisoara, Romania

Sharon Cooper, Oliver Cook and Natasha Loines
Durham University, UK

Marek Smolik
Faculty of Chemistry, Silesian University of Technology, Gliwice, Poland